SpringerBriefs in Space Development

Series Editor
Joseph N. Pelton, Jr.

More information about this series at
http://www.springer.com/series/10058

Scott Madry

Space Systems for Disaster Warning, Response, and Recovery

Scott Madry
Southern Hemisphere Program
International Space University
Chapel Hill, North Carolina
USA

ISSN 2191-8171   ISSN 2191-818X (electronic)
ISBN 978-1-4939-1512-5   ISBN 978-1-4939-1513-2 (eBook)
DOI 10.1007/978-1-4939-1513-2
Springer New York Heidelberg Dordrecht London

Library of Congress Control Number: 2014946078

© Scott Madry 2015

This work is subject to copyright. All rights are reserved by the Publisher, whether the whole or part of the material is concerned, specifically the rights of translation, reprinting, reuse of illustrations, recitation, broadcasting, reproduction on microfilms or in any other physical way, and transmission or information storage and retrieval, electronic adaptation, computer software, or by similar or dissimilar methodology now known or hereafter developed. Exempted from this legal reservation are brief excerpts in connection with reviews or scholarly analysis or material supplied specifically for the purpose of being entered and executed on a computer system, for exclusive use by the purchaser of the work. Duplication of this publication or parts thereof is permitted only under the provisions of the Copyright Law of the Publisher's location, in its current version, and permission for use must always be obtained from Springer. Permissions for use may be obtained through RightsLink at the Copyright Clearance Center. Violations are liable to prosecution under the respective Copyright Law.

The use of general descriptive names, registered names, trademarks, service marks, etc. in this publication does not imply, even in the absence of a specific statement, that such names are exempt from the relevant protective laws and regulations and therefore free for general use.

While the advice and information in this book are believed to be true and accurate at the date of publication, neither the authors nor the editors nor the publisher can accept any legal responsibility for any errors or omissions that may be made. The publisher makes no warranty, express or implied, with respect to the material contained herein.

Printed on acid-free paper

Springer is part of Springer Science+Business Media (www.springer.com)

This Springer book is published in collaboration with the International Space University. At its central campus in Strasbourg, France, and at various locations around the world, the ISU provides graduate-level training to the future leaders of the global space community. The university offers a 2-month Space Studies Program, a 5-week Southern Hemisphere Program, a 1-year Executive MBA and a 1-year Master's program related to space science, space engineering, systems engineering, space policy and law, business and management, and space and society.

These programs give international graduate students and young space professionals the opportunity to learn while solving complex problems in an intercultural environment. Since its founding in 1987, the International Space University has graduated more than 3,000 students from 100 countries, creating an international network of professionals and leaders. ISU faculty and lecturers from around the world have published hundreds of books and articles on space exploration, applications, science and development.

This book is dedicated to the men and women around the world, professionals and volunteers, who work in the disaster planning and response community at every level. Their dedication, sacrifice and skill are an inspiration, and they deserve the very best technology and tools that can be developed. I must also express my sincere appreciation to several people without whom this work could not have been possible. I want to thank Prof. Joe Pelton, my long time friend and colleague, and my wife Sarah and daughter Adrienne for their longtime support and help.

The superior man, when resting in safety, does not forget that danger may come. When in a state of security, he does not forget the possibility of ruin. When all is orderly, he does not forget that disorder may come. Thus his person is not endangered, and his States and all their clans are preserved.

Confucius (K'ung-fu-tzu) (551–479 B.C.)

Contents

1	**Introduction**...	1
	Why Write a 'Short Book' on Space Disaster Management?........	1
	How Can We Best Understand the Field of Space Systems and Disaster Management?................................	1
2	**Disaster Management and the Emergency Management Culture**..	5
	Modern Emergency Management............................	12
	Time Is the Enemy..	14
3	**Organizing for Disasters**...................................	17
	The U.S. Incident Command Structure (ICS)..................	17
	The Incident Team Response to a Disaster.....................	18
	Organizing the Response Under the National Incident Management System (NIMS)................................	20
	Emergency Operations Centers (EOCs).......................	25
	International Disaster Organization – The U.N. Cluster System......	38
	How Data Are Acquired, Processed, and Shared in a Disaster........	41
	Data Acquisition.....................................	41
	Data Processing, Analysis, and Synthesis...................	42
	Phases (Internal and External)...............................	42
	Disaster and Damage Assessment............................	43
	Daily Disaster Assessment...............................	44
	Weather Briefings......................................	44
	Disaster Data Dissemination..............................	44
	Action Plans..	44
	Situation Reports (SITREPS).............................	45
	Public Outreach, Press Relations, and Fundraising................	46

4	**Space Systems for Disaster Management**	47
	Satellite Telecommunications	48
	Satellite Telecommunications Systems Capabilities	49
	GEO: The Geostationary Telecommunications Satellites	51
	Fixed Services (FSS)	53
	Mobile Satellite Services (MSS)	54
	Broadcast Satellite Services (BSS)	55
	GEO Telecommunications Systems	56
	MEO Satellite Telecommunications Systems	59
	LEO Satellite Telecommunications Systems	60
	Iridium	60
	Globalstar	61
	Orbcom	62
	Satellite Telecommunications Disaster Applications and Issues	62
	The American Red Cross	64
	Satellite Telecommunications Emerging Technologies	64
	South America and Asia	65
	Unmanned Aerial Vehicles (UAVs) or High Altitude Platform Systems	65
5	**Space Remote Sensing Fundamentals and Disaster Applications**	67
	Geostationary Weather Satellite Systems	68
	Moderate Resolution Optical Satellite Systems	73
	The MODIS Sensors	77
	The NASA Tropical Rainfall Measuring Mission (TRMM)	78
	Landsat	78
	The Disaster Monitoring Constellation (DMC)	79
	High Resolution Optical Satellite Systems	80
	Developing Technologies	81
	Passive vs. Active Remote Sensing Systems	82
	Fundamentals of Active/RADAR Remote Sensing	84
	Aerial Imaging Systems and Unmanned Drones	88
	Developing Remote Sensing Technologies	89
6	**Precision Navigation and Timing Systems**	91
	PNT Systems	91
	Disaster Applications	94
	GPS In-Situ Networks	95
7	**Geographic Information Systems**	99
	Critical GIS Data Layers	99
	Open Source	101
	GIS Systems and Examples in Disaster Management	102
	U.S. National Weather Service Online Damage Assessment Toolkit	103

	Data Dissemination Systems	105
	Early Warning Systems	108
8	**Major International and Regional Players**	111
	The United Nations	111
	The U.N. Space-Based Platform for Information for Disaster Management and Emergency Response (UN-SPIDER)	111
	The U.N. Institute for Training and Research (UNITAR)	112
	The International Federation of the Red Cross Red Crescent Societies (RCRC)	112
	The Committee on Earth Observation Systems (CEOS)	112
	The CEOS Visualization Environment (COVE)	113
	The Group on Earth Observation (GEO) and the Group on Earth Observations System of Systems GEOSS	113
	FEWS Net, the Famine Early Warning Network III	114
	Asia Disaster Reduction Center (ADRC)	114
	Sentinal Asia	115
	The Pacific Disaster Center	115
9	**The Emerging World of Crowd Sourcing, Social Media, Citizen Science, and Remote Support Operations in Disasters**	117
	Crowd Sourcing	117
	Social Media	118
	Citizen Science and Remote Support Operations	121
	Major Players in the V&TC Community	121
10	**International Treaties, Non-binding Agreements, and Policy and Legal Issues**	123
	The International Charter on Space and Disasters	123
	The Tampere Convention on the Provision of Telecommunication Resources for Disaster Mitigation and Relief Operations	127
	International Non-binding Agreements	127
	Legal and International Policy Issues	127
11	**Future Directions and the Top Ten Things to Know About Space Systems and Disasters**	129
	Future Directions	129
	Top Ten Things to Know About Space and Disasters	131

Appendix A: Key Terms and Acronyms 133

Appendix B: Selected Bibliography 139

Appendix C: Selected Websites 143

About the Author 145

Chapter 1
Introduction

Why Write a 'Short Book' on Space Disaster Management?

Disasters occur every day on our planet, affecting millions of people each year and inflicting terrible suffering and economic loss. Space systems and our access to space technologies provide disaster managers with powerful capabilities that help to plan for, respond to, and minimize the impacts of disasters every day. Space systems have become a vital aspect of our global infrastructure, providing telecommunications, data, imagery, positioning and timing services effortlessly around the globe, but these are actually very complex systems to manage and use, and they are often the only such capabilities available immediately after a disaster has occurred, because they do not rely on ground services or infrastructure.

How space systems are used in disaster management, what capabilities are used for which activities, and where this is all going in the future is the topic of this book. If you are interested in how we plan for and organize our response to disasters around the world, then this book will be of interest, as it will explore the nature of disaster planning and response, the types of disaster that occur around the world, and how the disaster community is organized and structured. If you are interested in space technologies and applications, this book is also for you. Disaster management is a fascinating and complex subject with many different practical applications, all of which provide real benefits to society.

How Can We Best Understand the Field of Space Systems and Disaster Management?

The best way to get a sense of this field is to recognize that it is the interface of two very different domains. On one hand there is the very high technology field of space and space applications. On the other hand there is the very down-to-earth world of natural and manmade disasters that requires planning and training for, as well as immediate response to these catastrophic incidents.

Disaster management is a very complex, difficult, and dangerous activity. It is an arena where lives and property are at risk and decisions have tremendous consequences. Our access to space – and the many capabilities that our access to space provides to us – are an important part of the evolving toolkit now utilized by disaster managers. Yet these are very different fields of work and the interface between the two is often not effective. There are often misunderstandings – and even occasionally mistrust – on both sides of this ever more crucial partnership.

This book is meant to encourage all the different groups to better understand one other. Space professionals want to use their knowledge and capabilities to assist in disasters. Often these high-tech engineers and scientists have limited understanding about what is needed in the field and what can and cannot work effectively for disaster managers.

Disaster professionals are seeking better ways to operate and to incorporate advanced technologies into their vital work, but only when it makes sense. A new high-tech tool must prove itself to be more effective, practical, reliable and not excessively difficult to use in the field over techniques and technology that have worked in the past in a reliable manner. Newer and higher tech is not necessarily better. This is particularly true if the new device is more complex, heavier, more challenging to power, more difficult to use, not as reliable, and more expensive to train for and acquire.

It is important to grasp the quite different perspectives of space technologists and rescue professionals. One must begin by understanding that different standards of success and utility can and do apply. These two groups have a different understanding and regard for new technologies. The space technologist and the rescue worker are unique in terms of culture, process, and perspective in terms of effective disaster recovery and prevention.

Until these differences are understood it is difficult to improve the cooperative relationship and working interfaces between the two. The objective of this book is to define better ways to bring the amazing capabilities of space technologies into the daily activities and accepted practices of disaster responders around the world. This is a big challenge because the two groups do not really communicate well with each other. The differences are many and include different levels of technological development, training, financial resources, and understanding of needs and practical constraints as to what can and cannot work.

This book thus presents an overview of both worlds and the interface between the two. The goal is to provide sufficient context and detail so that anyone can understand, at an introductory level, what constitutes each domain, and how they can better work together – at each end of the process – to improve the important work of the disaster responders around the world.

Many people believe that disasters occur, and then one must leap into action to organize a recovery. Actually disasters imply a cycle that includes: (i) preparation; (ii) response; (iii) recovery, and (iv) mitigation. Then the cycle renews itself in the preparation stage in future planning and the mitigation processes that seek to avoid a future recurrence of the just past disaster. This proactive approach to disasters that uses the experience gained from one disaster to avoid or lessen the impact of a future disaster is key. It is in this disaster cycle that space systems can play a useful role at each of the four stages.

This book is not intended to be a highly technical treatise with extensive footnoting and detailed information about the complex technologies of space. Other books in the International Space University series provide information about telecommunications, remote sensing, weather and space navigation satellite technologies. Rather it introduces how space systems can play a vital role in disaster recovery and prevention in a wide range of ways.

The charts and graphs and case studies presented here are largely self-explanatory and for the most part do not require additional references or an advanced understanding of satellite technology applications. Those who find the ideas and materials presented in this book of particular interest and wish to pursue their studies in this field further should refer to the many Internet site references provided at the end of this book. This can be helpful in learning about the latest activities and technologies being developed in this exciting field around the world under the sponsorship of many nations, international agencies, disaster aid organizations and satellite operators. There are also cooperative efforts by those operating satellite networks for telecommunications, remote sensing, meteorology, and space positioning, navigation and timing.

Finally, for ease of reference, there is a list of key terms and of acronyms provided as an appendix at the end of the book. The chapters in this brief book are presented in what is hoped to be a logical order, but they can be for the most part read in any order one might choose, based on your own interests and priorities. Please enjoy this rather condensed book about a very important subject.

Chapter 2
Disaster Management and the Emergency Management Culture

Disasters are a fact of life, and this will always be the case. There are a wide variety of both natural and manmade disasters that occur around the world daily, and their impact can be both devastating and long-lasting. Natural disasters result from processes such as earthquakes, floods and cyclones, volcanoes, and even extreme solar storms and the impact of asteroids, while human-caused disasters result from societal or even individual actions. These can include epidemics, intentionally set forest fires, riots, toxic spills, and chemical, nuclear, and biological accidents or terrorist attacks. Some incidents are a combination of the two, such as the intentional sabotage of a dam that floods a large area. Some of these incidents occur with some amount of warning, such as tropical cyclones, but many, such as earthquakes, tsunamis, and tornados, occur suddenly, and can impact large areas with little or no prior warning.

No matter the cause, all major disasters require an immediate, comprehensive, and professional response. Major disasters create the most extreme stresses on a nation, its people, and government. As human population increases and the world's climate changes, we are seeing an increase in the intensity of disasters. Disaster planners and responders use a wide variety of technologies and tools to assist them in the chaos that surrounds any such incident. This book addresses the interface of space and space-related technologies and the disaster professional's procedures and toolkit.

What exactly is a disaster? Here are some basic definitions. An emergency can be defined as a series of natural or human incidents that endangers people, property and the natural and built environment. A house fire or serious auto accident is an emergency, but can be handled readily by the available resources within a community. A hazard refers to the physical characteristics that can create an emergency, such as living in a town that could suffer an earthquake or is in a dangerous location for landslides. Risk is defined as the potential or likelihood that a given emergency might occur, and risk management is the process of understanding, preparing for, and limiting risks as much as may be possible. Assessment is the process of analyzing a given situation, either before or during an incident. Preparedness is the process of planning, training and other activities that relate to dealing with emergencies and disasters. The vast majority of emergency management work is spent in preparedness. An incident is an occurrence, natural or manmade, that requires a response to protect life or property.

A disaster is an emergency that cannot be effectively managed using existing, local resources and implies a broad-scale incident that covers many people and affected infrastructure. Any time a community is overwhelmed by an incident, and it cannot effectively manage the situation using its existing resources, requiring a call for outside assistance, that is considered to be a disaster.

Disasters can be at many different scales and on many different timeframes. Emergency management is the profession and process of planning for, responding to, and minimizing the impact of emergencies and disasters. Emergency managers are the professionals in each community and nation who specialize in these activities, and are generally governmental employees considered to be civilian first responders, similar to and, in fact, often are police or fire fighters.

Response is the immediate set of actions taken to save lives and property in a disaster, including search and rescue, sheltering, emergency medical assistance, and more. Recovery is the longer-term returning the community back to normal, as much as possible, after the immediate response is over. Resilience defines how capable a community is to absorb these incidents without requiring outside assistance, and mitigation defines actions and planning to reduce the impact of the next incident, for example by moving people out of areas that repeatedly flood or enforcing stricter building codes in areas subject to earthquakes. So a disaster is a fundamental breakdown between humans, their environment and supporting technologies. When a community cannot adequately respond to an incident and requires outside assistance, then it is a disaster, and the emergency management community responds.

Disasters inflict terrible economic and human costs throughout our world. As just one example, the 1995 earthquake in Kobe, Japan, with a magnitude of 6.9, occurred in a major, highly developed urban area that was well built and very prepared for such an incident. It killed nearly 7,000 people outright and made over 600,000 people homeless, destroying over 150,000 buildings and damaging almost 200,000 other structures, many beyond repair. Infrastructure, roads, and utilities were devastated throughout a large area, and it took years for these to be rebuilt. The financial cost was an astounding US$114 billion, three times any previously recorded disaster. Sadly, this number has since been eclipsed by more recent events such as Hurricane Katrina in 2005 in the United States (US$125 billion) and the Japanese Sendai tsunami (US$235 billion) in 2011.

Other disasters impact much larger regions. The Indian Ocean great tsunami created massive destruction and loss of life in Indonesia, Malaysia, India, Sri Lanka, Thailand, and Myanmar with over 250,000 fatalities.

Although these are the largest and most well known recent disaster events, smaller regional and local disasters occur around the world on an almost daily basis. If your home and life are disrupted in a small flood or forest fire, your life is no less affected than if it was a huge event affecting thousands of people.

All disasters are ultimately local in their effect, and the impacts are deeply personal. Over the past decade, over 900,000 people were killed in disasters around the world, and over 2.6 billion were affected in some way. The economic losses were well over a half trillion U.S. dollars, with poor people and less developed countries being by far the most affected, particularly in terms of loss of life and ability to recover.

The following series of charts and data are derived from the OFDA/CRED International Disaster Database (http://www.emdat.be). This database, first created in 1988 by the World Health Organization's Collaborating Center for Research on the Epidemiology of Disasters (CRED), contains detailed information collected for over 16,000 disasters from 1900 to 2008. Over this period, and around the world, natural disasters made up 63 % of the collective events. Human-created incidents, rather regrettably, made up the remaining 37 %. Weather-related incidents, including cyclones and hurricanes, floods, and droughts are the most prevalent and deadly, followed by geological disasters such as earthquakes, landslides and tsunamis, and biological events (See Fig. 2.1).

Figure 2.1 shows that weather disasters are by far the most prevalent, with geological events second and biological third.

Figure 2.2 shows the total number of natural disasters reported from 1900 to 2007. Part of this trend represents the fact that many earlier incidents simply were not reported, but the sharply upward trend is still alarming. Other reasons for the increase in disasters is a function of a rapid rise in global population, increasing urbanization, and more and more reliance on modern infrastructure.

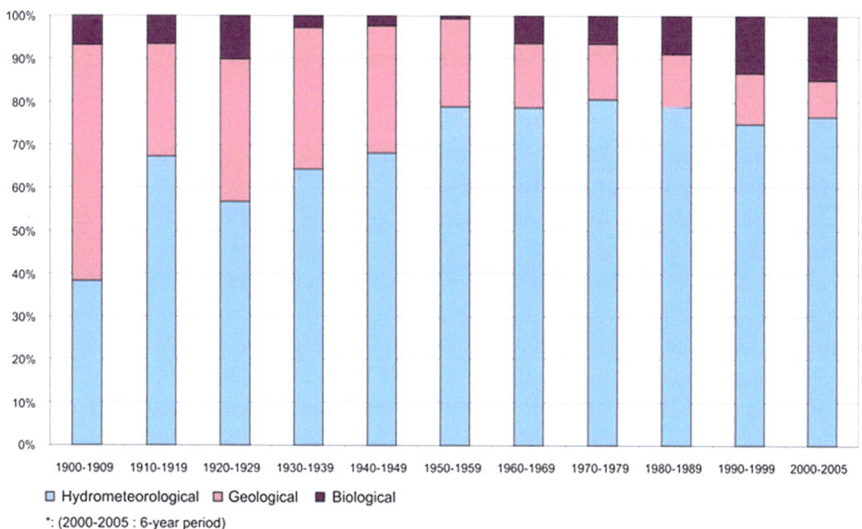

Fig. 2.1 World Health Organization statistics on global disaster events (Graphic courtesy of the World Health Organization)

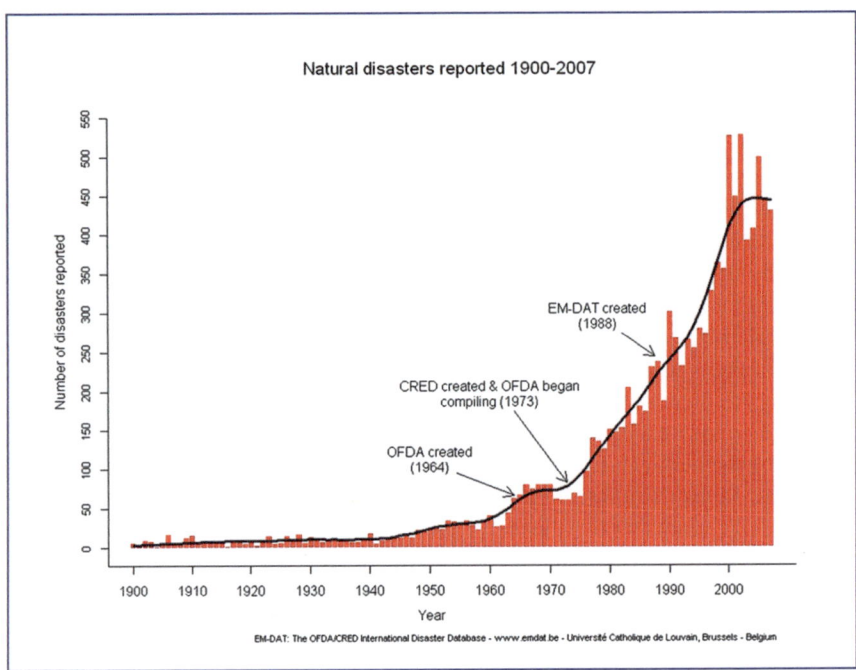

Fig. 2.2 Number of reported international disasters since 1900, ending in 2007 (Graphic courtesy of the World Health Organization)

Figure 2.3 shows the total number of people affected by disasters in the same time period, and again the world's growing population clearly puts more people at risk each time a major disaster strikes. The fact that many people, including many of the world's poor, live along the world's coastal zones makes them more vulnerable to many types of major disasters. Increased reliance on electrical power grids, modern transport and distribution systems, and urbanization often serves to increase the impact of a disaster.

Figure 2.4 shows the increasing estimated amounts of economic damage caused by these disaster incidents. The Kobe, Japan, earthquake of 1995 and Hurricane Katrina in the United States in 2005 stand out as extremely costly, due to their impact on the infrastructure of these two highly developed nations. Hurricane Katrina damage cost was estimated at over US$125 billion in 2005 dollars, while the Kobe earthquake damage estimate is about US$114 billion, or 2.3 % of Japan's GDP at the time – three times the cost of any previous recorded disaster. The 2011 Sendai tsunami is not shown, but it exceeded both at over US$235 billion – a truly staggering loss that involved the combination of a natural and a manmade disaster.

Figures 2.5 and 2.6 chart the overall increase in disasters, as well as the extreme increase in the impact of weather events around the world. Some of this may be due to lack of reporting of earlier storms, but the trend, as we said, is still alarming.

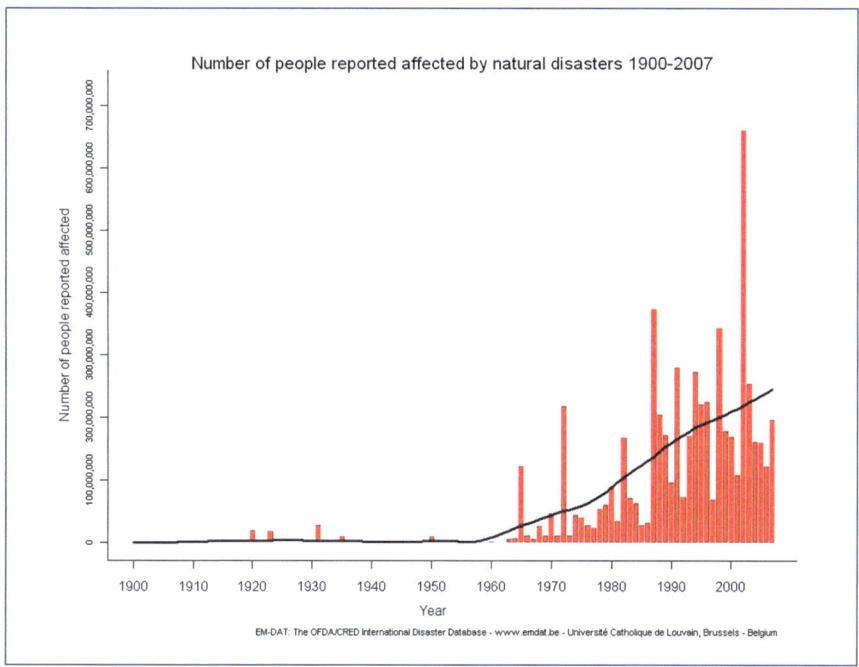

Fig. 2.3 Number of people affected by disasters from 1900 to 2007 (Graphic courtesy of the World Health Organization)

Figures 2.7 and 2.8 show the various types of natural disasters by continent. Africa is still plagued by biological disasters such as the Ebola virus. Thus 81.4 % of Africa's disasters have been in this arena. Europe, in contrast, has had very nearly 95 % of its disasters concentrated in floods and storm events.

Finally, Fig. 2.9 shows the number of deaths caused by disasters. The good news among these disaster statistics is that, even with the increasing impact and frequencies of these disasters, the total number of persons killed in these incidents has actually decreased. This result has been achieved even with the huge growth of global population and better reporting of disasters. This is largely because of the now common use of satellite weather and telecommunications and other capabilities to provide adequate early warning and evacuation for cyclones and hurricanes. Space systems are a key element in the improved response capabilities for all types of disasters.

Since improved space monitoring is the key element that has changed, it is perhaps useful to look a more recent data that covers the space age in greater detail.

Figure 2.10 shows the distribution of people killed in disasters more recently and also sorts the fatalities based on income levels. It shows the extreme impact of all types of disasters on the poor around the world. All types of disasters affect the poor disproportionately, and in all phases. They are less likely to receive warning and

10   2 Disaster Management and the Emergency Management Culture

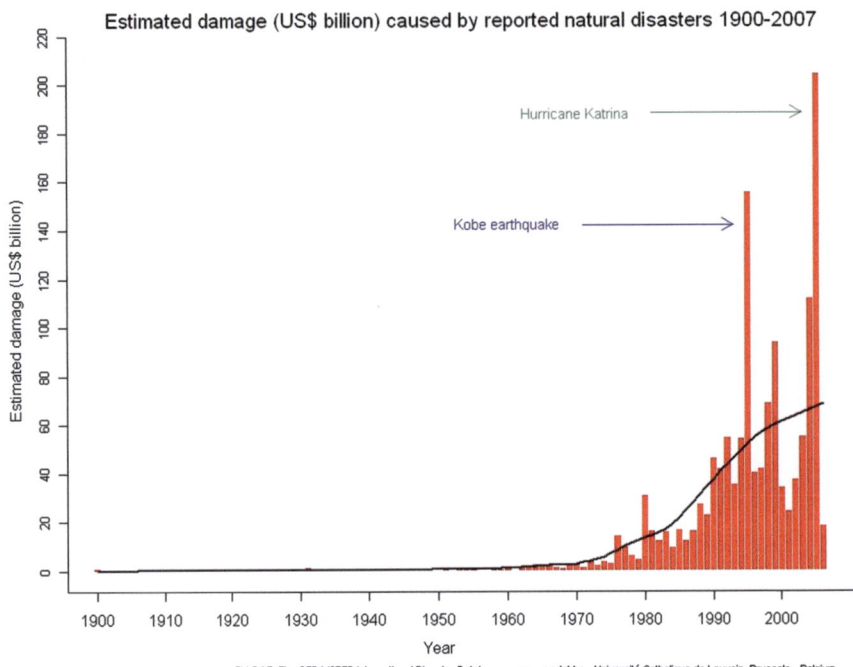

Fig. 2.4 Estimated amount of economic damage from global disasters from 1900 to 2007 (Graphic courtesy of the World Health Organization)

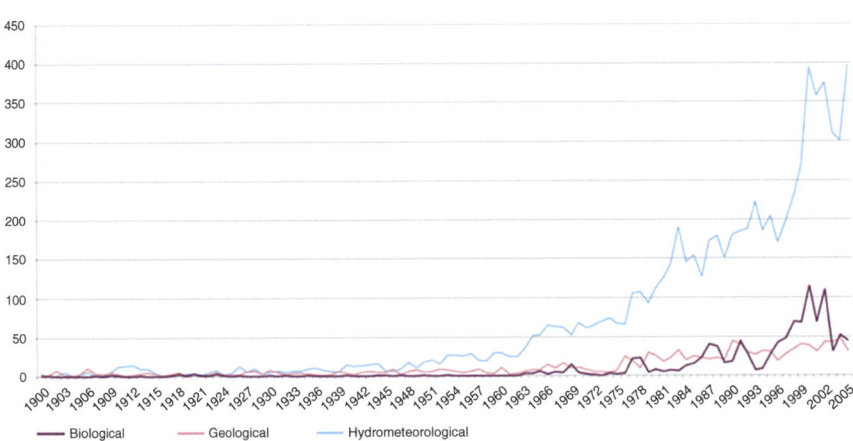

Fig. 2.5 Number of biological, geological, and flood/storm disaster events from 1900 to 2007 (Graphic courtesy of the World Health Organization)

2 Disaster Management and the Emergency Management Culture

	1900-1909	1910-1919	1920-1929	1930-1939	1940-1949	1950-1959	1960-1969	1970-1979	1980-1989	1990-1999	2000-2005	Total
Hydrometeorological	28	72	56	72	120	232	463	776	1498	2034	2135	7486
Geological	40	28	33	37	52	60	88	124	232	325	233	1252
Biological	5	7	10	3	4	2	37	64	170	361	420	1083
Total	73	107	99	112	176	294	588	964	1900	2720	2788	9821

Fig. 2.6 This table highlights the dramatic increase in disaster events since 1900 (Graphic courtesy of the World Health Organization)

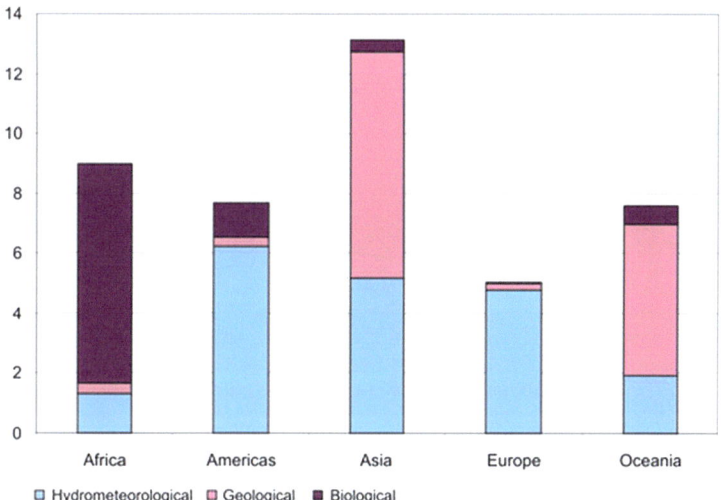

Fig. 2.7 Regional distribution of disaster events from 1900 to 2007 (Graphic courtesy of the World Health Organization)

	Hydrometeorological	Geological	Biological
Africa	1.30	0.37	7.31
Americas	6.23	0.31	1.13
Asia	5.19	7.54	0.39
Europe	4.77	0.23	0.03
Oceania	1.92	5.06	0.62

Fig. 2.8 Numerical breakdown of disaster events by region (Graphic courtesy of the World Health Organization)

evacuate, more likely to live in a vulnerable structure, more likely to die or become permanently displaced, and less likely to recover economically after a disaster.

Together, these charts show the extreme impact that disasters have on our world. The sheer numbers of people involved, and the vast amount of infrastructure that can be destroyed, underscore the vital importance of the work of the international disaster response community.

Who are these disaster relief and recover workers? How do they do their work?

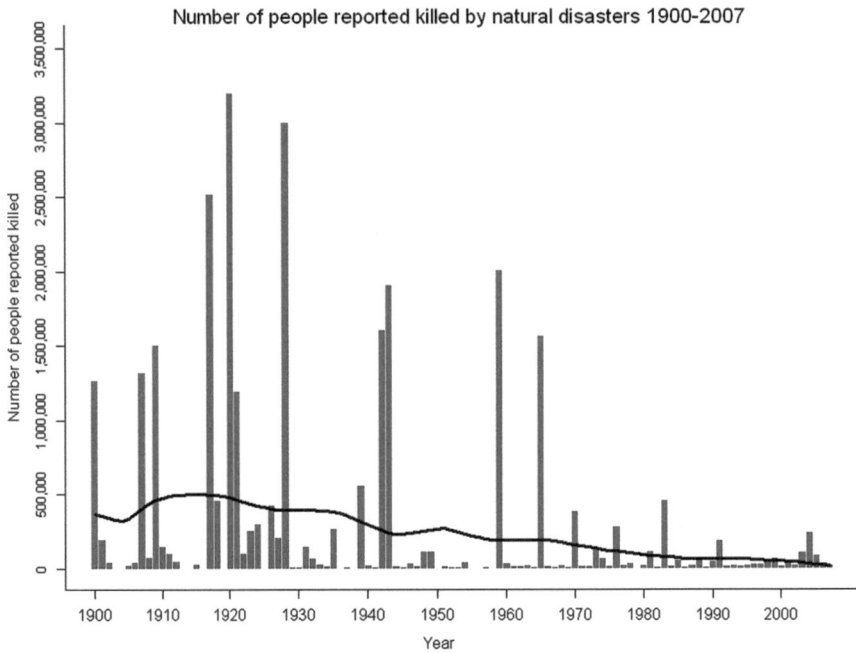

Fig. 2.9 Number of fatalities due to disaster from 1900 to 2007 (Graphic courtesy of the World Health Organization)

Modern Emergency Management

Emergency management spans a wide variety of activities, but focuses broadly on the reduction of vulnerability to hazards and preparing effective responses to all types and scales of disasters. One major goal is to promote and create safer and more resilient (less vulnerable) communities with the ability to better cope with natural and human-caused disasters. This is a challenging task, because of the wide variety of incidents that do and will continue to occur in the future. As the human population continues to grow and our footprint continues to expand over our world, we will feel these incidents ever more. Global climate change will also accelerate these impacts, especially in the coastal zones.

Once an incident occurs that exceeds the ability of a local community to effectively respond, then professional disaster managers, along with a wide variety of non-governmental organizations (NGOs) such as the Red Cross, voluntary aid organizations, local volunteers, military, and other entities respond to save lives and property. These professionals restore the community as much as possible to the condition before the disaster struck, and also seek ways to make the community more resilient in terms of the next disaster.

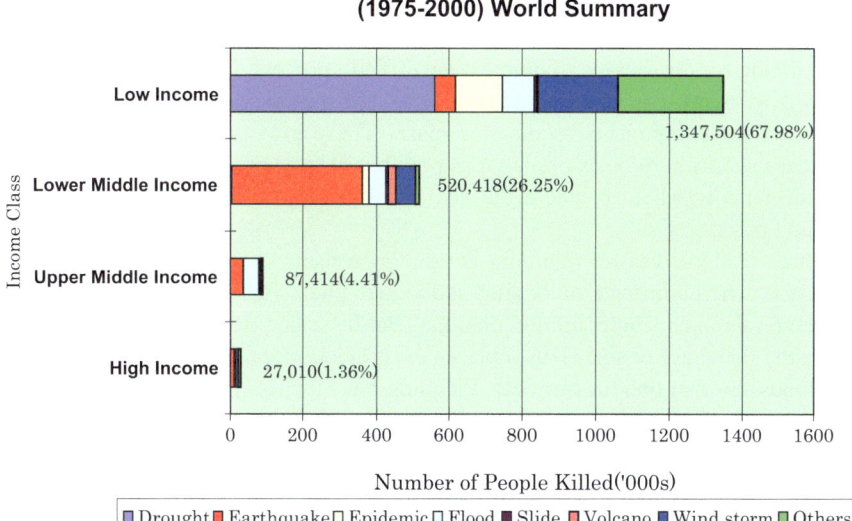

Fig. 2.10 Disaster fatalities between 1975 and 2000 grouped by income distribution (Graphic courtesy of the World Health Organization)

As noted in Chap. 1, there are generally thought to be four interconnected phases of the disaster management cycle. Most of the time is spent in the **preparedness** phase, where training, exercises, planning and educational activities for both the public and responders are conducted. Reading this book is part of preparedness. Once an incident occurs that exceeds a local community's ability to effectively respond, we enter the **response** phase, where immediate efforts are made to minimize the hazards, conduct search and rescue, open and manage shelters, provide public information, distribute food and supplies, provide medical assistance, and generally deal with the immediate situation as quickly and effectively as possible. After the initial response phase, we enter the **recovery** period, when efforts are made to return the community, environment, and economy back to normal.

Depending on the nature and scale of the incident, this can take a short or an extended period of time. For example, after a major earthquake or cyclone, the recovery phase can take years and cost billions. The final phase is called **mitigation**, where lessons learned are discussed, the response is evaluated, and steps are taken to minimize the effects of the next disaster. This can include, for example, buying land in flood-prone areas and perhaps turning them into parks and recreation areas so that when the area floods again there will no buildings or residents there. Another part of mitigation is to strengthen zoning or building construction codes so that structures can better withstand earthquakes or severe storms. These are only two examples of possible longer-term steps to minimize the effects of the next incident. This is because emergency professionals know that it is not *if* but *when* the next disaster will strike. The cycle thus begins again with the preparedness process. The disaster cycle never ends, and, as stated before, the majority of time is spent in preparedness.

Planning for disasters is a complex and difficult process. Each local jurisdiction, as well as many private organizations and aid agencies, creates and annually updates their comprehensive disaster plans. These groups or jurisdictions hold exercises and train to be as prepared as possible. It is a complex problem, because we never know when or where the next disaster may strike. Preparedness is an ongoing process.

Much modern disaster planning is now done using an 'all hazards/all scales' approach where, instead of having individual plans for each possible type of incident, there is a single, unified plan that can be scaled up to meet any scenario, no matter the cause or size. This replaces the previous approach of having one plan for floods, another one for fires, etc. Planning must include detailed 'how to' plans that are to be followed, including contact lists, which entity or organization is to fulfill what roles, inventories of critical infrastructure, demographic information about who is living in the area, maps, details of previous disasters, the likelihood of reoccurrence, and more.

Time Is the Enemy

In the end, all disasters are local incidents, no matter how severe or widespread the damage. The initial response is conducted by local volunteers and available resources, and after the immediate response is completed, it is the local people and resources that must pick up the pieces and rebuild their lives. In responding to a major disaster, time is the real enemy. Most casualties who survive arrive at hospitals or aid stations within the first few hours, and local volunteers and the first groups of responders rescue most survivors within the first few days. For the severely injured or people trapped in rubble, the chances of survival past these first critical days drop off dramatically.

Time is the enemy, and getting accurate information about the scope, extent, and impact of the disaster is critical to creating and orchestrating an effective disaster response and recovery effort. This is where space technologies can and should play an important role. The overall context for space technologies for disaster response lies in the domain of situational awareness and command and control. Once a disaster occurs, an initial assessment needs to be conducted as soon as possible to determine the scope, geographical distribution, and scale of the incident. Situational awareness means being able to accurately determine what has happened, what is happening now, and what will come next, all in order to plan and coordinate the most effective response possible with the resources available.

Getting the *right* people and resources to the *right* place at the *right* time is the essence of the command and control aspect of the disaster response. It is also the greatest challenge. This occurs at both a tactical or daily, localized level, as well as on a larger, strategic scale. The main objective often becomes to get the right information to the public, to emergency responders, and to governmental decision makers, and also briefing responders in time to make a difference. Nothing is less useful than information that does not arrive in time or that is sent to the wrong people who do not need it. It is critical for first responders in an emergency to avoid acting on incorrect information and instead get the right information in a timely manner.

Command and control involves the infrastructure, tools and procedures that allow you to harness the resources that you have and to communicate and guide the response effort. It includes the operational decision-making environment and infrastructure that is used to manage an incident – providing the who, what, when, where, and how of the response. Each day is a different set of interrelated problems. You have a given amount of uniquely trained responders, various vehicles and resources such as food, medical supplies, search and rescue teams, fuel, radios, etc. You need to have information on where the most urgent requirements are. It is critical to have information on the current situation such as where roads are closed, where power is or is not available, which airports and hospitals are open, etc. This is a constantly changing matrix that requires frequent updates. Key updates to the matrix might include the arrival of new resources, or that some roads or bridges are cleared. Other key updates might include information about shelters being opened or closed or power lost or restored. Making the right decisions at the right time saves lives and property. Key information that accurately updates in a timely way a disaster management information matrix represents the heart of the disaster response effort.

When a single regional or national entity is providing the response this is still complex, but a simpler problem in disaster management and communications. Often, in a major national and international response, there are many governmental entities, NGOs and other players that are collecting data and providing resources, as well as seeking access to timely and accurate data from many parts of the response community. Space systems that provide telecommunications, remote sensing, mapping, weather, space navigation, etc., can be a key part of this process.

Although progress has been made in incorporating space and other advanced technologies into the disaster response process, there is still much work that needs to be done. Many disaster incidents, especially in less developed parts of the world, still do not effectively incorporate these powerful capabilities. Why is this the case, and how can this be improved? Part of the answer lies in the nature of the professional emergency management and response community.

Professional emergency management (EM) professionals around the world are a very disciplined and structured community. This community is closely aligned, as would be expected, with the professional police, fire and military communities, and operates with similar hierarchies, procedures, and logistics. The EM community is very structured and conservative, with specific protocols, checklists, training, and defined, modular capabilities, asset, and roles. This is due to both the scope and difficulty of disaster response, and also due to the fact that lives and property are at stake. This is a very serious business.

EM organization budgets are always under pressure, as over 90 % of the time is spent in the preparedness phase, and it looks like nothing is happening, so why do we need to have all this expensive equipment and capabilities? But once a major event occurs, the response is often quickly and publicly criticized by the affected community, and through the press, for not responding quickly enough. Certainly, each failure or perceived lack of effective response can be and likely will be highlighted publicly via the media and the Internet.

These factors, and the difficulty of conducting operations such as urban search or mounting rescue operations in remote and severely impacted locations, tend to create a very conservative and highly structured organizational response mind-set. The EM community, generally speaking, is not accustomed to ad hoc response or to 'on the fly' incorporation of new techniques and approaches. The traditional mode of operation is to 'train like you respond, and respond like you train.' Emergency responders spend much time each year training with sophisticated and often very expensive tools. It follows almost absolutely that the tools they use must work, as needed, each and every time, and that the staff can learn to operate them effectively.

Every part of the emergency management system must fulfill specific and integrated roles in the organizational structure and work flow. Standardized training is important, and technologies and tools have to be able to be used by the rank-and-file EM workers and be deployed into difficult working environments where external support is limited or nonexistent. Therefore, training is a vital component in the use of any equipment or technology. Being able to integrate advanced technologies into the standard training and working protocols of the EM community is an important aspect of getting the EM community to adopt space technologies and data.

Much more needs to be done to effectively incorporate advanced space technologies into the regular training, deployment and use cycle that is provided to emergency responders. The reasons for this better connection have been indicated already. There needs to be truly systematic training and use of any new technology tools during realistic simulation exercises. This type of systematic preparation, especially for satellite-related tools and information, underscores this basic message. The providers of satellite systems technology and information need to be more closely tied to the routine training and applications process for the emergency response providers. Showing up unannounced in a disaster zone with new, high-tech equipment is just not the way to successfully incorporate advanced technologies into the response. One positive development in recent years, in this respect, is that a number of satellite providers have created separate units that are tasked with fulfilling the needs of military, defense, and emergency response organizations.

Chapter 3
Organizing for Disasters

Disasters occur all over the world, and the response is clearly influenced by the local culture, level of development, historical context, and the national readiness to respond. The key to the response is the structure of the emergency response officials. Around the world there tends to be two main types of emergency response management structures. One is known as the incident command structure that has largely been developed in the United States. The other approach is that of the U.N. cluster organization. Both of these approaches and their current derivative management concepts are described in this chapter. First we will start with the U.S. "incident command structure," or ICS, and then proceed to discuss the U.N.'s cluster management system.

The U.S. Incident Command Structure (ICS)

In the United States, each county or parish has a professional emergency manager, who is responsible for planning and preparing for disasters in their jurisdiction. At the state level, each state has an emergency management organization that has the responsibility at that level. At the federal level, FEMA, the Federal Emergency Management Agency, is the lead governmental unit. Nevertheless every federal agency has specific responsibilities in the event of a major disaster. There are many organizations, such as the Red Cross, Salvation Army, Care International, and others, who also are involved in disaster planning. During this extended preparedness phase, much work is done, including holding training exercises, developing training staff, updating plans and maps, creating memorandums-of-cooperation, and more.

Planning for and responding to disasters includes a variety of planning assumptions. These include the assumption that incidents will occur without warning at the worst possible time, such as the Boxing Day tsunami, when many responders were on vacation. Further, in terms of worst case planning, there has to be an assumption that essential infrastructure such as key hospitals or even emergency management facilities may not be useable. This worst case type of planning thus leads to other types of preparations. This has led to a requirement for disaster response facilities.

Further, all disaster response entities are required to maintain a continuity of operations (COOP) plan. This COOP plan sets forth how they will maintain operational capability if their key facilities are destroyed or not available, and if their key staff cannot respond. Key response personnel are often unavailable, either due to their being out of town or they are taking care of their own families or are injured or killed in the incident themselves. Nearby locations that would normally be relied upon to provide support through mutual aid agreements may be impacted themselves and be unavailable to assist (or seeking aid from you), and critical infrastructure such as roads, power, and telephone lines, Internet connectivity, water and sewage lines, telephone lines, cellular sites, etc., can be disrupted. Planning has to be for the worst-case scenario, and the assumption has to be that every incident is different, and brings with it a unique set of challenges. Even the same type of incident in the same location a few years later will have its own unique challenges and difficulties.

In a major incident, the general requirement is that all responders must be self-sufficient, bringing all resources including food, shelter, medical capabilities, etc., to the incident, and must be self-reliant for the duration of their time in the affected zone. This is often not the case, and unprepared responders or 'self-deployed' volunteers often put a strain on the response effort that exceeds any benefit they may provide. Professionals never self-deploy, but work within a highly structured process of deploying specifically trained skills in a coordinated process. Disaster planning today is generally 'all hazards/all scales,' where there is one single, integrated plan that includes the response plans for all types and scales of events and which defines the specific roles played by each organization. Detailed maps showing the demographics of the area are kept and updated, as it is vital to know who is being affected in a given area, their economic status, special language needs, and many more aspects as well.

The Incident Team Response to a Disaster

Once a major disaster occurs, the incident response swings into action. As in a military campaign, no matter how much training and preparedness have been done, each response is unique and will contain unanticipated challenges and surprises. The most carefully laid plans quickly have to be adapted, and the need for accurate and timely information of the incident is key to getting the right resources to the right locations in time to be effective. If the incident exceeds local capabilities, then mutual aid requests must be made to adjoining jurisdictions or to a higher level of government. Space technologies can and do play important roles in this aspect of the response. Coordination between the various responders is difficult, and the response is sometimes half-jokingly referred to as the 'second disaster.' Different organizations use different organizational structures, logistics, communications and staffing structures, and coordinating these into a single, integrated response is always a challenge in the best case.

In major international responses, we often see a lack of timely and useful information, a lack of coordination, and political, organizational, and personality conflicts that hinder the response between organizations. In addition there are turf battles and tensions, conflicting agendas, a lack of cultural sensitivity, inappropriate aid (such as winter clothing donations arriving in the tropics), seeking religious or political gain in return for assistance, spreading rumors and more. In short, a major international disaster response can be an exercise in managing chaos.

We also see well-intentioned but unqualified volunteers clogging the system and consuming scarce resources. Responders who are not self-sufficient end up requiring scarce supplies, power and housing. Disaster tourists who simply want to experience the situation also complicate the response and recovery. There can even be bizarre and pathological behavior, such as persons with false credentials seeking to 'play doctor' or otherwise misrepresent their expertise. Other complications can be groups providing aid in return for political or religious goals, criminals stealing and looting, and aberrant behavior. Looting and other criminal behavior are often a major problem, including theft of response equipment and supplies, and even attacks on disaster responders.

Managing the response can be as difficult as managing the disaster, and this is all made worse by the number of individual groups responding. Problems that can emerge in large-scale disaster responses include a general lack of integrated communications, lack of electrical power and or data links, extreme heat or cold in the disaster area, stress, crime, poor hygiene, and more. Managing a major disaster is an extremely difficult, complex, and continually evolving scenario. Each day brings a new set of challenges, needs, and opportunities. Planning the next day's activities is a continually changing matrix of need, altered conditions, and available resources.

What is needed is one integrated response, where the total combination of resources available on a given day are properly allocated to the specific needs and priorities of that day. But this is very difficult to manage in reality. The larger and more isolated the incident the more difficult it is to manage the various organizations, logistics, demands, and needs. Coordination of large incidents on a daily basis is very important but understandably hard. Effective management of technology can both be a major help but can also a major problem, in that it requires valuable electrical power, fuel, staff, and security resources.

One of the realities of emergency response is that each incident and response is unique. Even a flood or earthquake in the same region a decade apart will have very different characteristics. But there are common requirements and needs in all major disasters. There must be an overall command and leadership element to coordinate the response. Within this, there must also be several individual units to provide needed capabilities and services. These often include planning, logistics, mass care and sheltering, administration and finance, public affairs, security, water and sanitation, and so on.

There is a need for a logistics group to get the needed resources where they need to be, a staffing and planning unit to provide the needed skills, and human resources as required.

Organizing the Response Under the National Incident Management System (NIMS)

In the United States, the overarching framework for major disaster management is the National Incident Management System (NIMS). Within this is the U.S. Incident Command System (ICS), which is how the disaster response process is structured and managed. As already stated, all disasters are local, and the response begins and ends with the local responsible officials.

The organizational structure of the EM community reflects both the importance and difficulty of their job. Each nation is responsible for the safety and welfare of its citizens, and each nation organizes this effort in its own way. Many nation-states operate their national EM capability as a part of their public safety, fire or military organizational structure, while others operate independent EM organizations.

In the highly developed industrialized nations, EM is a well-funded professional process, but the situation is much less provided for in the rest of the world. In the United States, there developed a disaster response and recovery organizational structure called the Incident Command System that is also used in many nations around the world, including Canada, the UK, Australia, and New Zealand (in slightly different forms).

The ICS system is an all hazards/all scales system of response organization, meaning that it can scale up to any required size and can respond to any type or combination of incidents, and then shrink back down as the incident evolves. The system was first developed by the multi-organizational forest fire fighting community in California in 1968, where multiple levels of organizations and overlapping jurisdictions had to find a way to effectively work together without having to constantly decide who was in charge depending on the location or players in a given situation. The ICS system has evolved into a very robust and capable means of organizing large, multi-party responses that can meet any type of need.

Some of the essential features of ICS are that there is a single command structure, and while command can be transferred across organizations and levels of government, there is only one person in charge at any time, called the incident commander (IC). The IC can come from any one of several jurisdictions or organizations as needed. In incidents involving multiple jurisdictions, the ICS unified command structure allows different agencies to work together efficiently, and to share resources. There is an understanding that a common terminology will be used to describe functions, facilities, positions, and responsibilities. Communications are always 'in the clear,' rather than codes such as '10-codes,' which can have different meanings to different groups. Confusion as to the meaning of a code can be a serious danger when someone might understand a code to mean a grave illness condition where in fact there might be a fire or dangerous gas leak.

Clear and open communications are a major aspect of ICS, and all communications are thus sent using a common language that employs non-coded and standardized terms. There is an orderly chain of command, as well as a unity of command, so that everyone involved in the response has a clear role and a defined supervisor, and managers at all levels are directly responsible for those under them, no matter what organization they come from, and, in turn, everyone knows to whom they report.

Another key component is management by objectives, where plans and procedures are established to support defined goals, and metrics are in place to determine effectiveness. There is a modular organization structure that can expand or shrink, depending on the situation. An incident action plan (IAP) provides a clear means to communicate the specific operational goals, and these plans do not exceed 24 to 48 h in length, in order to remain relevant and focused on the immediate situation.

Another key element of the ICS is referred to as 'span of control.' Within the ICS structure, any individual supervisor only has between 3 and 7 subordinates, usually only 5. When the available or needed staff exceeds this, then another level of organization is added. This keeps the management structure lean and effective, and no individual supervisor manages more than 5 to 7 people and often less. All locations are designated using a common set of terms, such as command posts, staging areas, and so on. Also, there is always comprehensive resource management, regardless of the source of the resources. This process is used to maintain an accurate and up-to-date picture of resource availability and is always known throughout all the various participating organizations. It is this span of control and resource management system that allows the ICS to scale up to a very large organization in the event of a very large disaster and then scale back down as needed.

The ICS system develops and uses a common communications plan and seeks interoperable communications architectures and systems. In a number of instances, this interoperable communications is provided via fixed or mobile satellite communications networks. The ICS system specifically maintains a process for information and intelligence management, and has a dedicated structure for the gathering, analyzing, sharing and use of information and intelligence. Satellite imagery, weather maps, GPS data, and GIS mapping are key components of this activity. Thus ICS is quite compatible with the use of various types of satellite applications.

The incident commander operates out of the incident command post, or ICP, where the overall response is directed. There is only one ICP at any one time, but it can change locations as needed. It can simply be a single fire truck, a mobile command truck or bus, or be located in a trailer or tent. The ICP is usually identified by a flashing green light and is given a specific icon for any map. It is located near the incident, but in a safe zone outside of danger. The ICP logo is a square divided along the diagonal into green and white triangles.

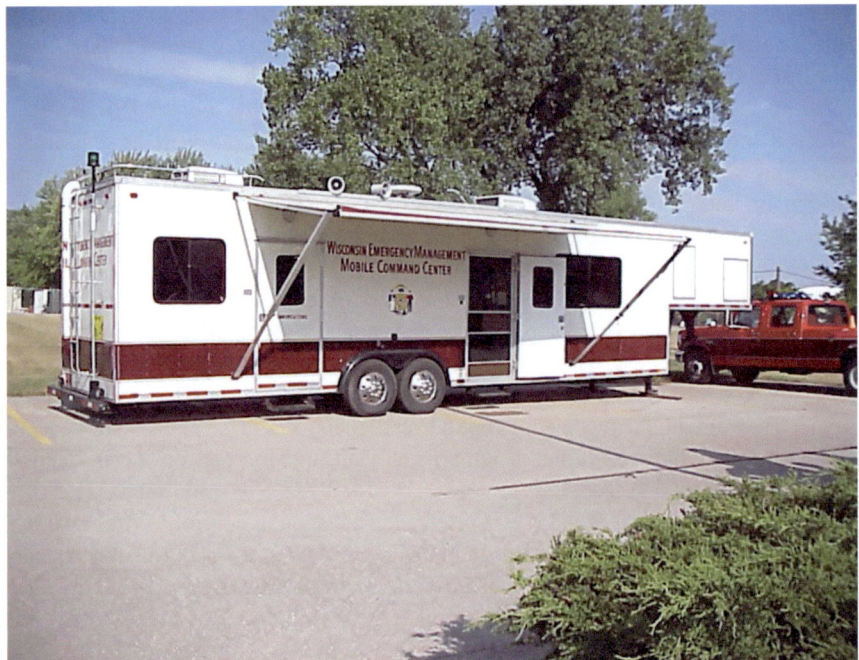

Fig. 3.1 Mobile command center (Graphic provided by the State of Wisconsin Emergency Management)

Mobile command posts are designed to provide sophisticated communications and computing resources to the incident commander in the field. Figure 3.1 shows an example, a mobile command center for the Wisconsin Emergency Management Agency. Note the green light at top left.

In the United States, the ICS is an integral part of the National Response Framework (NRF), where the ICS structure is used to directly manage the 'point of the spear' of the response. The NRF describes the larger structuring of resources and provides the overall context where all aspects of the nation's resources can be harnessed to support a major disaster response. The 2013 version incorporates a series of 15 emergency support functions (ESFs), which bundle together and manage resources into functional areas of responsibility. Within each ESF there will always be a lead agency or organization with overall responsibility for coordinating that activity.

There are 15 ESFs that provide the specific response capabilities in that area. The command organization structure leads the overall response, as shown in Fig. 3.2.

- ESF 1 Transportation
- ESF 2 Communications
- ESF 3 Public Works and Engineering
- ESF 4 Firefighting
- ESF 5 Information and Planning

- ESF 6 Mass Care, Emergency Assistance, temporary Housing and Human Services
- ESF 7 Logistics
- ESF 8 Public Health and Medical Services
- ESF 9 Search and Rescue
- ESF 10 Oil and Hazardous Materials
- ESF 11 Agriculture and Natural Resources
- ESF 12 Energy
- ESF 13 Public Safety and Security
- ESF 14 Long-Term Community Recovery
- ESF 15 External Affairs
  (http://www.fema.gov/media-library/assets/documents/32183?id=7354)

Each ESF has a lead federal agency that is responsible for the overall management of that aspect of the response, but many different federal, state, and local entities, along with aid organizations and others are key players in each ESF. For example, the American Red Cross plays a major role in ESF 6 by providing sheltering, mass feeding, damage assessment, and other activities. ESF 5 is Information and Planning, which collects, analyzes, processes, and disseminates information about a potential or actual incident and conducts planning activities to facilitate the overall activities in providing assistance to the entire response community. This is where satellite imagery, GPS and GIS data and maps, are developed, and where these specialized skills and expertise reside for a major disaster response. These skills also reside in all the other ESFs and at all levels, including federal, state, local, and others, but all coordination is done within ESF 5. Planning makes each incident action plan (IAP) and briefs the IC and staff at the start of each IAP cycle, usually 24 or 48 h.

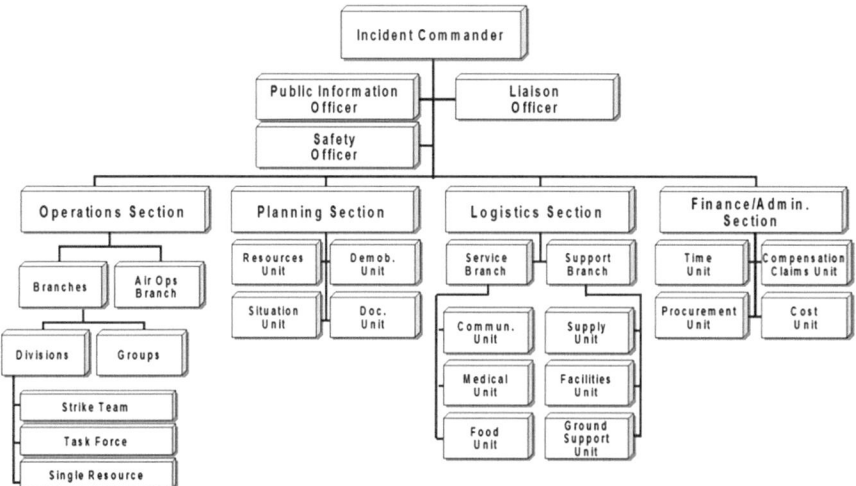

Fig. 3.2 The incident command system (Graphic courtesy of FEMA)

There is only one incident commander at a given time, and all responders ultimately report to the IC, regardless of their organization or affiliation. Within the IC's staff are the liaison officer, who deals with the various governmental and organizational liaisons, the public information officer, who communicates with the public through the press, and the safety officer, who is responsible for the overall safety of all activities and staff. The safety officer or staff can stop any action or person if they determine that safety is endangered. The Operations ESF carries out the actual response according to the current incident action plan, and Logistics provides the material for all aspects of the effort. Finance and Administration manages the people and money to make the response happen.

The ICS system is designed as an all-hazards and all-scales structure. It is designed to expand and/or contract as needed throughout an incident. For example, the first fire or police officer on a scene would become the initial IC but may quickly pass this role over to a supervisor or to another as the incident evolves. There can be a small staff or a major response can number in the thousands. Flexibility is key, and the response grows to meet the need, and then shrinks back as the situation evolves and finally ends. All participants must understand the ICS system, and FEMA provides online training, through hundreds of courses for emergency responders.

The details of this training and all aspects of the FEMA Emergency Management Institute can be found at their website, http://training.fema.gov. All courses are free, and any space technologists interested in assisting the disaster community should become familiar with the ICS by taking the appropriate online courses. This would show your basic knowledge of the ICS process and demonstrate that you are serious about working within the EM community.

Within the Planning staff is located the Situation Unit, and this is where the data analysis, mapping and GIS work is done, and this is where satellite imagery, GPS, social media, and other spatial data are acquired, processed, and delivered to specific users. The Situation Unit collects, processes, and organizes all information, prepares situational reports (SITREPS), and develops projections and forecasts of future events, including changes in the weather and other relevant information. The Planning Section Chief manages the staff, and they collect and manage all operational data, prepare the IAP, compile and display information, manage specialized data such as weather forecasts, and create and disseminate other maps and information for use throughout the response. Technical specialists, with specialized knowledge and expertise, work in the Planning section, or may be assigned elsewhere as needed. These include meteorologists, satellite image analysts, GIS and GPS mapping specialists, cartographers, environmental impact specialists, and others as needed.

Within the Planning section, the collection, analysis and sharing of intelligence are major activities. These focus on three incident intelligence areas: situational status, resource status, and anticipated incident status (such as changing weather forecasts and future plans). All of these together are used for incident management situational awareness and decision-making at all levels of the response. The overall goal of the Planning section is to evaluate the situation, develop incident objectives,

Emergency Operations Centers (EOCs)

select appropriate strategies, share the right information in the right format throughout the response, and decide which resources should be used to achieve the objectives in the safest, most efficient, and most cost-effective manner. Then Operations carries out the plan with the resources provided by Logistics and Finance and Administration.

The ICS Form 201 is the incident briefing form, usually referred to simply as the 201. This is a standardized, four-section document that is usually produced in four pages. The goal is to quickly capture a synopsis of the vital incident information. This is always developed whenever there is a command transition (a new IC takes charge). It is important to note that the very first thing on the first page of the IC is a map of the situation, showing the value and importance of spatial data. A blank copy of the current 201 is shown below in Fig. 3.3.

The ICS 209 form is the Incident Status Summary, and collects the basic incident decision support data. It is the standard situational report for administrators and executives. These would include government officials off-site that are not directly involved in the immediate response. See Fig. 3.4 below.

The ICS Form 205 is the Incident Radio Communications Plan that provides information on all radio frequency allocations at all levels. Information and telecommunications resources are detailed, and this would include satellite communications capabilities, mobile radio, and narrowband and broadband Internet resources. This function that operates under ESF 7 – Logistics – manages, tests, and installs all communications equipment, operates the incident communications center, and acquires, tests, and maintains all such equipment. A specific incident communications plan develops frequency inventories, frequency-use agreements, and interagency radio caches. The communications plan establishes networks for command, tactical, support, and air units and ensures that codes are not used. As noted earlier, all communications are accomplished using a clear spoken message, in order to avoid misunderstandings. This unit also enforces the clear communications mandate. This means that no codes (such as code 19) are used because different organizations have different codes meaning different things. There are usually several different communications nets, including command, tactical, support, air-to-ground, and air-to-air as needed. A major goal is to avoid frequency overlap.

Emergency Operations Centers (EOCs)

The ICS is all about the operational and field levels, where the actual response to a major disaster is coordinated, and where the 'battle' is fought. But there are several additional levels and organizational structures that are equally important, and all communications systems can and do use space technologies. The most important of these is the Emergency Operations Center (EOC). The EOC is the place where each level of government or each aid organization coordinates its disaster activities at a

1. Incident Name	2. Prepared by: (name)	INCIDENT
	Date:	BRIEFING
	Time:	ICS 201-CG

(include sketch, showing the total area of operations, the incident site/area, overflight results, trajectories, impacted shorelines, or other graphics depicting situational and response status)

Situation:

Fig. 3.3 (continued)

1. Incident Name	2. Prepared by: (name)	INCIDENT
	Date:	BRIEFING
	Time:	ICS 201-CG

5. Initial Response Objectives, Current Actions, Planned Actions	

Fig. 3.3 (continued)

1. Incident Name	2. Prepared by: (name)	INCIDENT
	Date:	BRIEFING
	Time:	ICS 201-CG

Fig. 3.3 (continued)

level removed from the actual ICS. In the United States, each county, state, and most federal agencies and organizations involved in EM activities operate an EOC. Some, like the FEMA or National Red Cross facilities, are 24/7, highly secure facilities that are always monitoring events and are constantly ready to begin coordinating activities should disaster strike. Others at the state and local level are 'stood up' or staffed on an as-needed basis when an incident occurs, but are not in constant operation. Smaller EOCs at the local level can simply be a conference room converted into this use when needed, but most states have very elaborate, secure, and expensive facilities with large screen displays, extensive computer resources, and even food, sleeping, and shower facilities for a 24/7 response to extended incidents (Fig. 3.5).

The EOC is one step removed from the actual Incident Command headquarters, and the EOC is where, for example, elected political officials and county government managers assess the situation and coordinate information and available resources. The EOC is the central location from which governments at each level provide interagency coordination and executive decision making in support of the

1. Incident Name	2. Prepared by: (name)	INCIDENT
	Date:	BRIEFING
	Time:	ICS 201-CG

6. Current Organization (fill in additional appropriate organization)

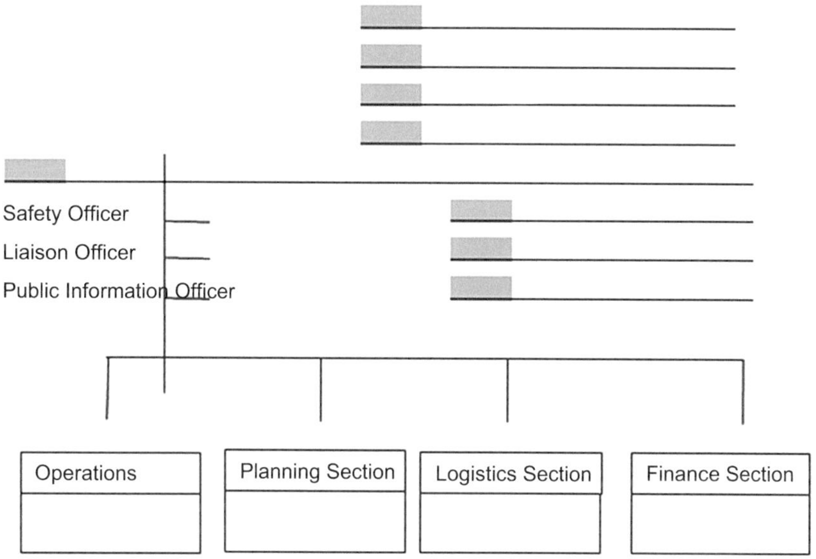

Safety Officer
Liaison Officer
Public Information Officer

Operations	Planning Section	Logistics Section	Finance Section

Fig. 3.3 (continued)

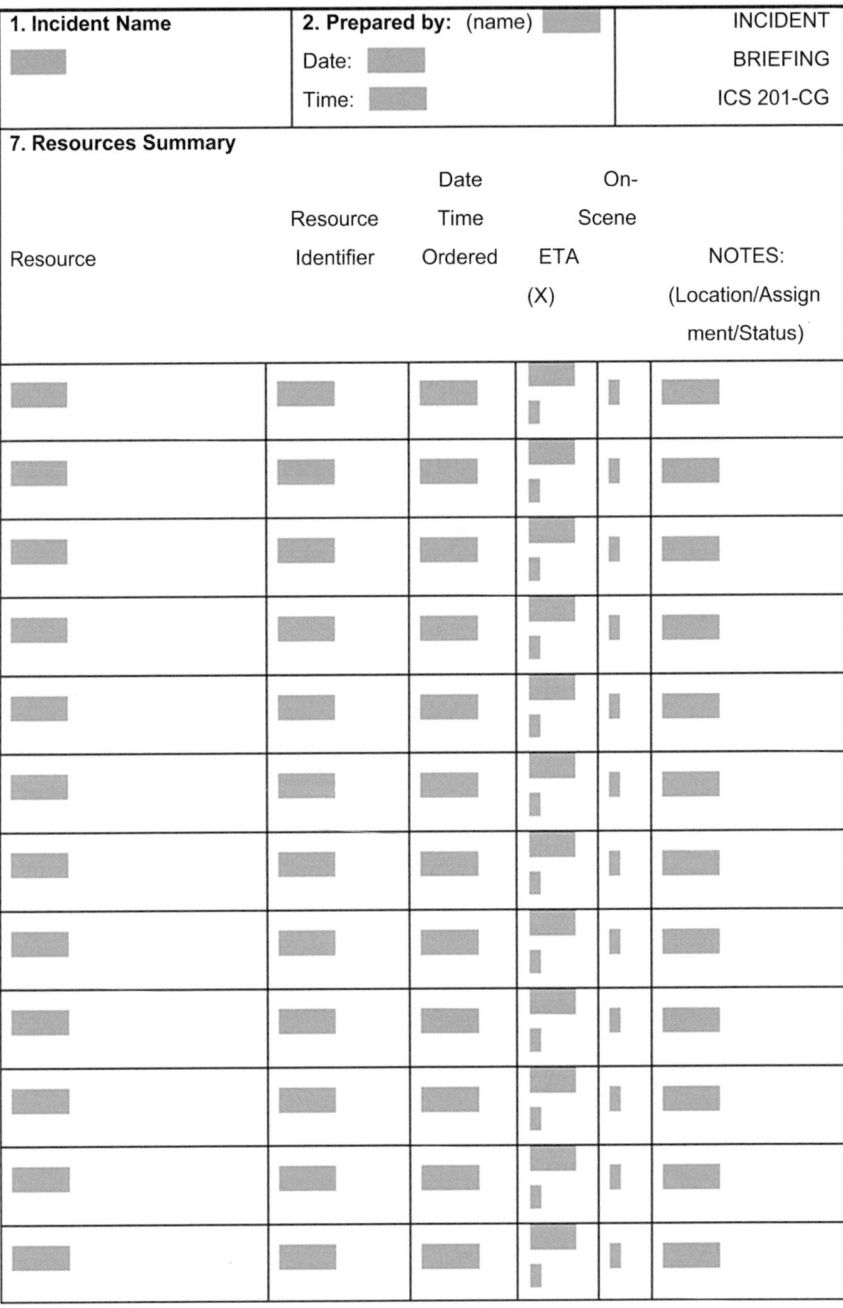

Fig. 3.3 (continued)

1. Incident Name	2. Prepared by: (name)			INCIDENT BRIEFING ICS 201-CG
	Date:			
	Time:			

Fig. 3.3 (continued)

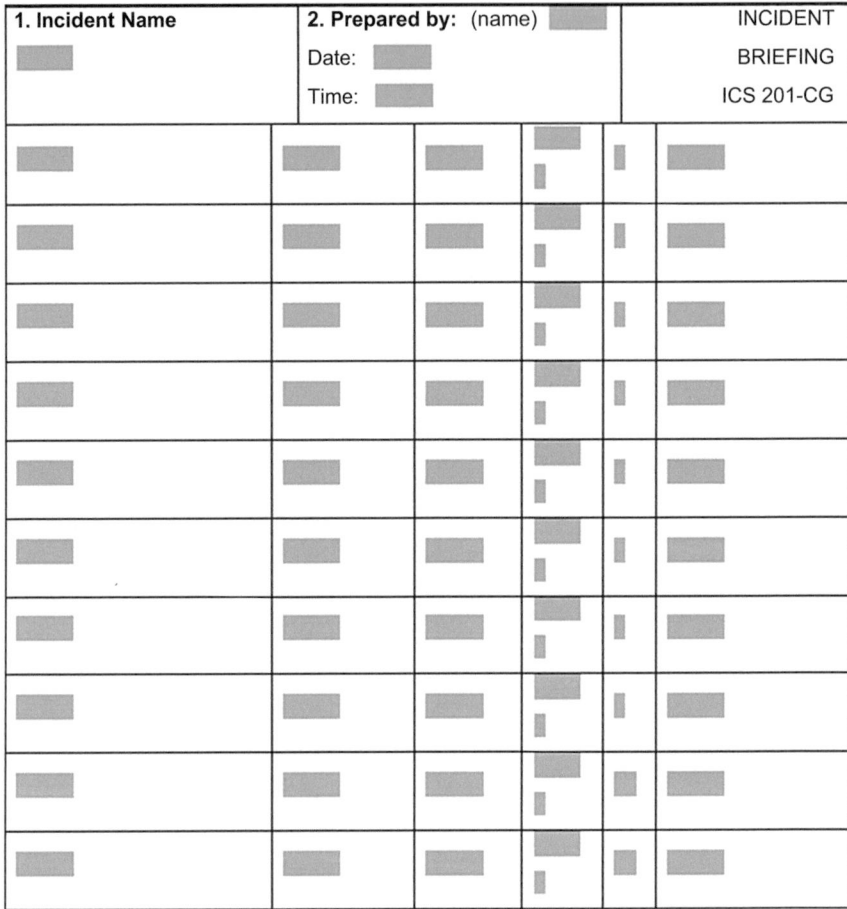

Fig. 3.3 ICA 201 incident form (Courtesy of FEMA)

incident response. EOCs exist at all levels, including, in the United States, at the county, state, and federal level, as well as for non-governmental agencies (NGOs) such as the Red Cross.

The EOC does not command the on-scene response, but carries out a higher-level coordination through information collection and evaluation, priority setting and resources management. Thus the EOC allows tactical decisions to be made by the incident commander and the command staff on the scene. EOCs are usually located in a secure facility with controlled access, and can also serve the role of continuity of operations, previously referred to as COOP, where elected officials can maintain governmental responsibilities if their normal locations have been impacted by the disaster.

1: Today's Date	2: Today's Time	3: Initial | Update | Final | |		4: Incident Number	5: Incident Name	
6: Incident Kind (WF – Full Suppression) (WF – Point or Zone Protection) (WF – Monitor/Confine/Contain)		7: Start Date Time	8: Cause	9: Incident Commander	10: Incident Command Organization	11: State-Unit
12: County	13: Latitude and Longitude Lat: Long: Ownership at origin:		14: Short Location Description (in reference to nearest town – i.e. 8 miles south of Show Low, AZ):			
15: Size/Area Involved (i.e. Acres)	16: % Contained or MMA	17: Expected Containment Date:	18: Line to Build (indicate chains, feet, meters, or miles)	19: Estimated Costs to Date:	20: Declared Controlled Date: Time:	
21: Injuries this Reporting Period:	22: Injuries to Date:	23: Fatalities	24: Structure Information			
			Type of Structure	# Threatened	# Damaged	# Destroyed
25: Threat to Human Life/Safety: Evacuation(s) in progress ---- ____ No evacuation(s) imminent -- ____ Potential future threat -------- ____ No likely threat --------------- ____			Residence			
			Commercial Property			
			Outbuilding/Other			

Fig. 3.4 (continued)

26: Projected incident movement / spread in 12, 24, 48 and 72 hour time frames: 12 hours: 24 hours: 48 hours: 72 hours:
27: Values at Risk: include communities, critical infrastructure, natural, and cultural resources in 12, 24, 48, and 72 hour time frames: 12 hours: 24 hours: 48 hours: 72 hours:
28: Critical Resource Needs (amount, type, kind, and number of operational periods in priority order in 12, 24,   48 and 72 hour time frames): **ex. 3 CRW1 (4); 1 HEL1 (5);** 12 hours: 24 hours: 48 hours: 72 hours:
29: Major problems and concerns (control problems, social/political/economic concerns or impacts, etc.) Relate critical resources needs identified above to the Incident Action Plan.

30: Current Weather for Current Operational Period Peak Gusts (mph): Max Temperature: Wind Direction: Min Relative Humidity:	31: Fuels/Materials Involved: (Insert primary Fire Behavior Fuel Model – i.e. 1 Short Grass (1 Foot))
32: Today's observed fire behavior (leave blank for non-fire events):	

Fig. 3.4 (continued)

33: Significant events today (closures, evacuations, significant progress made, etc.):			
34. Forecasted Weather for next Operational Period Wind Speed (mph): Temperature: Wind Direction: Relative Humidity:	34: Estimated Control Date: Time:	36: Projected Final Size:	37: Estimated Final Cost:
38: Actions planned for next operational period:			
39: For fire incidents, describe resistance to control in terms of:			
1. Growth Potential (Low, Medium, High, Extreme) -			
2. Difficulty of Terrain (Low, Medium, High, Extreme) -			
40: Given the current constraints, when will the chosen management strategy succeed?			
41: Projected Demobe Start Date: Time:			
42: Remarks:			

Fig. 3.4 (continued)

	43: Committed Resources											
Agency	CRW1	CRW2	HEL 1	HEL 2	HEL 3	ENGS	DOZR	WTDR	OVHD	Camp Crews	Total Personnel
	SRT	SRT	SR	SR	SR	SRT	SRT	SR	SR		
Total											
46: Cooperating and Assisting Agencies Not Listed Above:											
Approval Information											
47: Prepared by:	48: Approved by:	49: Sent to: Submission Date: Time:							by: Submission		

Fig. 3.4 ICS Form 209 – incident status summary (Graphic courtesy of FEMA)

Organization of the EOC is often organized by the Emergency Support Function (ESF) system. At a county level, the town planner, police, fire, public health, and public works managers would coordinate their efforts together, each acting in the role of an ESF lead. At the state level, a State EOC can be a very large facility with literally hundreds of people on call to represent every aspect of state government, along with NGOs and others, to coordinate their response efforts across the state. Facilities are available for the governor and staff, and a media and press room is used to communicate with the population through the press. Maps and satellite images, provided by the ICS Planning staff, are often used.

Emergency Operations Centers (EOCs)

Fig. 3.5 An Emergency Operations Center (EOC) in action (Image courtesy the City of Nashville, Tennessee. http://www.nashville.gov/Mayors-Office-of-Emergency-Management/Operations/Emergency-Operations-Center-[EOC].aspx)

A response is first handled at the local level, and it is only if local counties cannot manage a situation would they call upon the state authorities to assist. Once a local jurisdiction's resources are exceeded, then a request for state-level federal assistance is made. Once this is exceeded, a request for federal assistance is made. This is done by the governor of the state, by signing a state disaster declaration, which officially requests federal assistance. Multiple county EOCs in a state would be in constant contact with their state EOC. These local and state EOCs are one part of a larger system of multi-agency coordination. EOCs are both users and creators of spatial information and often include a dedicated GIS and mapping component.

There are other important users of space data in a response, including the following:

The Joint Operations Center (JOC) is an additional facility allowing multiple agencies to co-locate staff for liaison purposes, in addition to the various existing levels of EOCs.

The Joint Information Center (JIC) falls within ESF-15, external affairs. The JIC is established to coordinate public information activities between all of the players in a larger response, including all levels of government, NGOs, and others. It is often co-located with an EOC, so that elected officials can participate in the press briefings. The JIC is the central point for public affairs and media access and press briefings, and most organizations involved in the response will have public affairs representatives located there. The JIC is a major user of visuals, including

GIS maps, satellite imagery, websites, and video, as these are vital in explaining the situation to the public through the press.

A Joint Field Office, or JFO, may be established in a large disaster. When a disaster in the US is of such a magnitude that it requires participation by several federal departments and agencies, a Joint Field Office (JFO) is established near the incident, where senior federal representatives, called the JFO Coordination Group, provide strategic coordination of federal incident management activities. This is separate from any EOC, but has communications and liaison coordination with the EOCs. A temporary *Regional Response Coordination Center (RRCC)* can be activated by FEMA to coordinate the federal response to a major event until a JFO can be established. The functions are then absorbed into the JFO once it is established and operational.

International Disaster Organization – The U.N. Cluster System

International disaster responses become even more complex and difficult with more players and more complex logistics and communications requirements. The United Nations, which leads major international disaster responses, uses a 'cluster' system, with 11 individual clusters or sectors of the response activity, rather than the 15 ICS emergency support functions used in the ICS system. This was established in 1991 by the General Assembly Resolution 46/182 and has been updated and amended by the U.N. resolution under the name of Humanitarian Reform of 2005.

Each cluster is led by a designated U.N. agency or other entity, such as the International Red Cross and Red Crescent Society, with clear points of contact and accountability. The clusters are based on a sector or service provided in the disaster response, including protection, camp coordination and management, water sanitation and hygiene, health, emergency shelter, nutrition, emergency telecommunications, logistics, early recovery, education, and agriculture. (See Fig. 3.6 below.) The cluster system has been adopted widely around the world to deliver humanitarian assistance. The first use was in the 2005 Pakistan earthquake, and it has evolved and been in use ever since.

The U.N. Office for the Coordination of Humanitarian Affairs (OCHA) works with various United Nations and other organizations to coordinate and organize the field response. In the field, OCHA supports the humanitarian coordinator (HC), who is the individual appointed as the leader of the response and facilitates inter-cluster coordination. The response is led in the field by the HCs or the resident coordinators (RCs), who are experienced and trained disaster managers.

Each cluster has its own core responsibilities, including supporting service delivery within its area of responsibility. It keeps the HC informed and plans its own activities and monitors and reports on its work. Information management should link all the clusters together, but one problem with the U.N. cluster approach

International Disaster Organization – The U.N. Cluster System

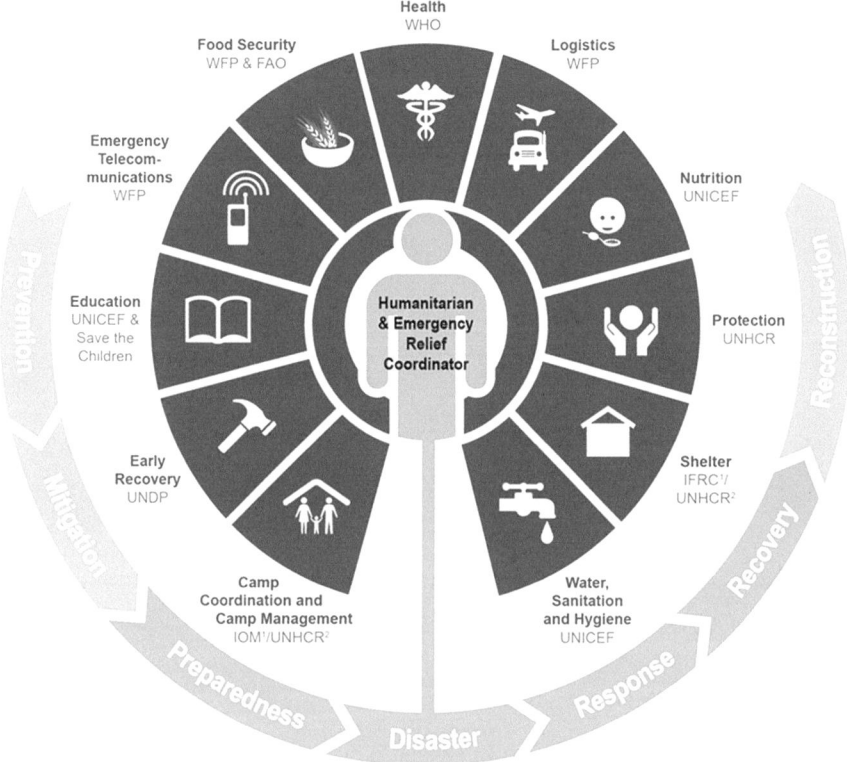

Fig. 3.6 The U.N. cluster disaster response organizational structure (Graphic courtesy of OCHA)

(and ICS as well) is that it is very easy for data and information acquired within each cluster to be 'stove-piped' within a cluster and not adequately shared to the larger response in a timely manner. The cluster approach structure is shown in Fig. 3.7.

The emergency relief coordinator leads the Inter-Agency Standing Committee, and is the undersecretary general for humanitarian affairs. The ERC appoints a humanitarian coordinator and ensures overall activities. The HC is accountable to the ERC and leads the overall response effort. The humanitarian country team (HCT) is the operational decision-making forum, led by the HC, and is composed of representatives from the United Nations, involved NGOs, and the Red Cross/Red Crescent Movement. It is responsible for strategic-level decisions.

At the national response level, the lead organizations of each cluster are directly accountable to the HC for ensuring coordination and relief activities within its cluster, and for coordination within the larger response. Clusters can be activated or deactivated as needed, allowing the response to match the evolving situation on the ground. Coordination between clusters is a constant challenge, and it is led at the strategic level by the HC and at the operational level by the cluster coordinators. There is an inter-cluster coordination forum (ICCF), which meets as a group of all

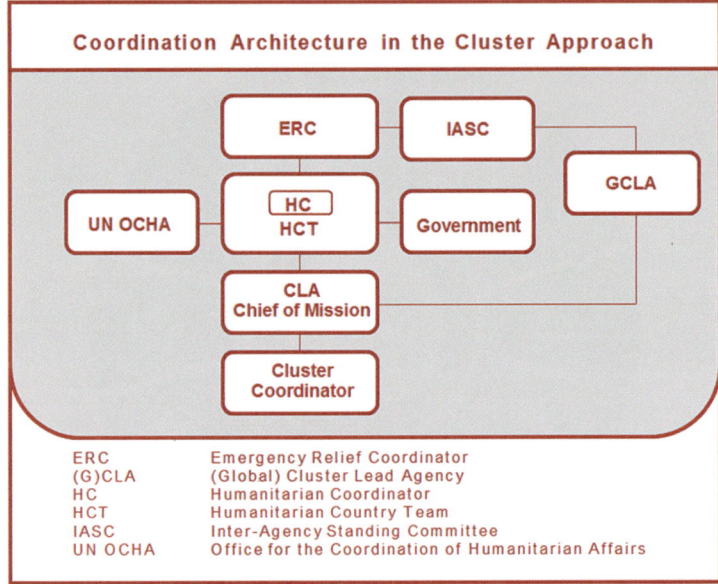

Fig. 3.7 The U.N. cluster disaster response coordination architecture (Graphic courtesy of OCHA)

clusters and also as small groups of related clusters as needed at the operational level.

Each cluster has six shared core functions:

1. Supporting service delivery
2. Informing strategic decision-making
3. Planning and strategy development
4. Advocacy
5. Monitoring and reporting
6. Contingency planning/preparedness/capacity building

A fundamental aspect of the cluster approach is one of shared leadership, where cluster lead agencies seek to share leadership between the United Nations and the Red Cross/Red Crescent/NGOs and others involved, in order to coordinate the response. The cluster approach seeks to support a needs-based response and avoid overlap and duplication. Each cluster knows its responsibilities and has a single point of contact within the overall response to address issues as they arise. A major aspect is also the inclusion of the affected communities in the response, providing access to decision makers and those with access to resources. For more information on the U.N. cluster approach, see the IASC Reference Module for Cluster Coordination at the Country Level: https://www.humanitarianresponse.info/system/files/documents/files/iasc-coordination-reference%20module-en_0.pdf.

How Data Are Acquired, Processed, and Shared in a Disaster

Acquiring, processing, and sharing the right data with the right people at the right time is the heart of disaster situational awareness and command and control. In a rapidly unfolding disaster situation, this is a very dynamic and complex process that can be a source of much difficulty. When it is done well, the entire disaster response benefits, and when it is done poorly, everything is more difficult. The entire disaster response is driven by the information available, so accurate and timely data acquisition, processing, and dissemination are vital. This is complicated by the fact that there are many sources of raw data, and many individual organizations, and each ESF or cluster has its own specific data needs and schedules. There is also much 'stove piping' of data and many structural problems in sharing information between organizations.

Disaster information management is intended to provide vital information for the decision-making process and to support relief planning and delivery. Likewise there is a need for critical information about the extent of damage, the impact of the incident and the current and future weather situation. Finally, there is a vital need to provide information for the media and public. Matching needs against capabilities is vital and continuing. This overall process is referred to as disaster assessment and consists of a continuing process of data acquisition, data processing, synthesis, analysis, and data dissemination.

Data Acquisition

Data on the current and future situation are continually acquired from multiple sources, including data from satellites, GPS, and GIS. It should be clear that not all of this information comes from space. There are many sources of information, including local media, the Internet, local governments, field reports, affected individuals, utilities and other participants in the response. Satellite-based information and information from stored GIS data systems are a valuable source of information, but are just a part of the overall information flow. This information collection process is continual throughout the incident. Sometimes the problem is the flow of too much information and not knowing which information source is reliable, accurate, and up to date.

The Red Cross refers to the several categories of needed data as the 'essential elements of information' (EEI), and these elements include a comprehensive set of information bits about the status of the situation in 15 categories:

1. The geographical extent of the incident
2. Hazard specific information on the incident
3. Jurisdictional and political boundaries of the incident

4. The demographics of the impacted area
5. Status of the emergency response and emergency operations centers
6. Status of communications and Internet
7. Status of transportation
8. Status of critical facilities
9. Status of electricity
10. Status of water and sewer/sanitation
11. Status of natural gas
12. Socio-economic/political impacts
13. Hazards, toxic, and radiological issues
14. Historical patterns and information
15. Weather and other environmental status and concerns

One of the significant challenges we face is that all of the above elements are constantly changing, and it is often difficult to track them all in a disaster situation.

Data Processing, Analysis, and Synthesis

A vital aspect of this is the analysis and synthesis of data. We need to turn raw data from multiple sources into targeted information that is tailored for specific end users, and getting that information to them in time to be useful. We cannot flood the managers of the response with tons of raw data, and so a vital role is the proper processing, synthesis, and delivery of the right data. The only thing that is worse than not having the right information in a disaster response is to be drowning in raw data that you cannot make sense of, or having data that is wrong.

Phases (Internal and External)

Once a disaster occurs, the first requirement is referred to as the initial impact assessment. This is the first understanding of the type, scope, and size of the incident. What has happened? What are the geographical boundaries? Who has been affected? What is the current situation? And what will the situation be tomorrow? Attention is focused on identifying what resources and trained responders are going to be required in the initial surge of the response. Once the response is initiated, a series of preliminary assessments are done, again more focused on getting the right trained responders, equipment, and aid to where it is needed and to identify problems as they arise.

Disaster and Damage Assessment

The process of disaster assessment (DA) allows immediate and recurring assessment of the impact of the disaster. The DA measures needs against available response capacity. It identifies resource needs, including financial needs, and provides information to the planning group for upcoming or anticipated needs. There is a vital need immediately after an incident to determine accurately the extent and nature of what happened. Several items included in this assessment are the type, extent, and severity of the disaster. The initial DA tries to identify the geographical area affected and the impact on transportation infrastructure such as roads, bridges, and airports. It also seeks to evaluate the status of critical sites such as power grids and hospitals, the status of local responders, and the number people affected and their demographics. This is followed by an assessment of the likely need for shelters and food, and what the situation will likely be in the immediate future. Is the incident over? What will the weather be like tomorrow? All these are vital to know as soon as possible for several reasons.

The main purpose of the assessment is to allow emergency managers to shape and begin delivery of the correct response. How many of what type of specialists skills and equipment will be required? How can responders get into the area? Where will they sleep? How long will it be until the situation is stabilized? Is the incident over, or is it continuing? All of these data are needed in order to get the right people and relief into the affected area to begin the response. A variety of data are synthesized into this initial picture, coming from a variety of sources, including media, the Internet, various utilities, governmental agencies, and others. Once the initial assessment is completed, a number of maps, graphs, charts, and a variety of other output data are generated for both internal and external use.

As soon as the response is underway, several preliminary damage assessments (PDAs) are conducted by various players, in order to define as quickly as possible the actual extent of damage to housing and infrastructure. These are often specific to the individual groups that are responding and their particular role and need. These concerns cover such issues as the condition of roads by the highway department, the extent of the power outages by individual utilities, status of water by utilities, etc. In the United States, the Red Cross conducts a PDA within the first hours of an incident to determine how many people were affected in order to plan how many shelters will be needed and where, what feeding will be required, and which skills and responders will be sent to the affected area. One problem is the consistent sharing and updating of this information among the various players. This lack of information sharing is sometimes called stove-piping. This means information travels vertically within a particular unit, but it does not travel horizontally to other groups.

Daily Disaster Assessment

Once the response is underway, each day presents a new mix of challenges. Roads are cleared, new resources arrive, some power is restored, shelters open and close. The day's plan needs to be made to match the day's needs with available resources, and this is driven by analysis of incoming data.

Weather Briefings

The continual assessment of weather is a vital aspect of the disaster response, and involves the collection and sharing of data concerning the current and projected weather. This includes specific information to ensure the safety of the responders as well as those in the impacted area. Maps and weather reports are generated as needed, and are used to ensure sufficient water and other resources both for the affected communities and for the responders.

Disaster Data Dissemination

Once the latest data are acquired and analyzed, we have to share the data in appropriate ways. This means sharing the right data, both internally within all the responding organizations and more broadly to decision makers and the public. There are many needs and ways to share information in a disaster. Often critical information does not get to the right person or unit or arrives too late, and maps based on satellite imagery are an important part of this aspect of the response.

Action Plans

Every response, whether under the ICS, U.N. cluster, or another organizational structure, relies on an action plan. Under the ICS system, the incident action plan (IAP) is the formal document that outlines the specific and immediate goals of the response, and this is referred to as control objectives in the ICS system. This is the 'battle plan' for the response that is defined by the incident command and contains the goals, tactics, and allocation of resources. The plan is usually updated every 24 to 48 h, and also includes the safety and logistics plans, maps, radio frequencies, and other critical information needed to accomplish the objectives.

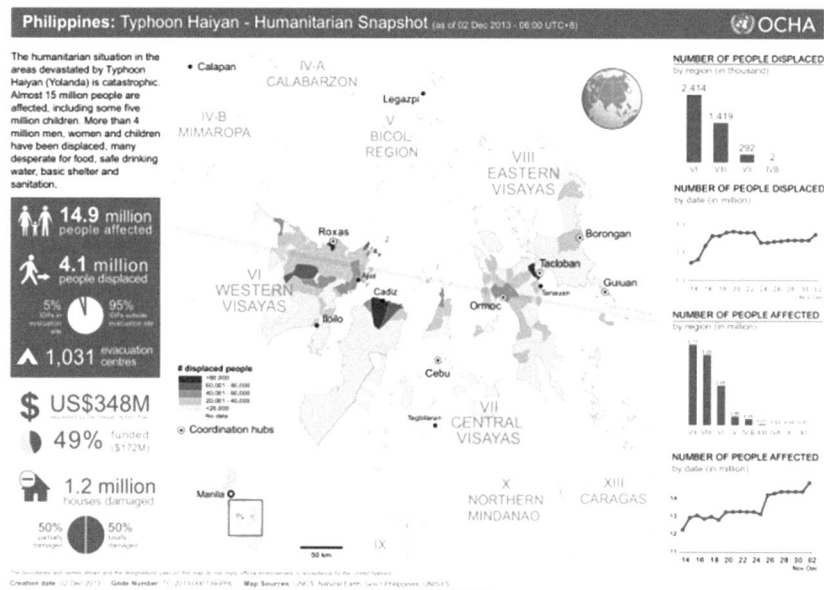

Fig. 3.8 Example of a daily humanitarian snapshot of the typhoon Haiyan in the Philippines (Graphic courtesy of OCHA)

Situation Reports (SITREPS)

The primary internal means of sharing updated information within the response team is in the SITREP, a standardized format report that provides a consistent, structured listing of periodic situation reports. Within the ICS system, each state usually has its own variation on the same overall theme, and a major federal response also uses a standardized SITREP form. It is often limited to four pages, with a map of the response situation at the top of the first page. Within U.N.-led international responses. for major international incidents, there is a standard OCHA SITREP format used. These are issued in the emergent phase of an operation, and usually only for major incidents, but they are also used routinely by responders in the field as well as management, public affairs, etc. There are several other standardized reporting forms, including what is called a "humanitarian snapshot," which is a simple info-graphic that is useful for quick overviews, and which includes simple graphics, maps, and text overviews of the current situation.

In Fig. 3.8, is a humanitarian snapshot used for the typhoon Hainan in the Philippines in 2013.

Public Outreach, Press Relations, and Fundraising

In the U.N. system there is what is called ReliefWeb (http://reliefweb.int). This is a dedicated website managed by OCHA that is designed to be a one-source website to disseminate all types of data in a disaster. In order to maintain timeliness, ReliefWeb is updated around the clock by a U.N. staff in New York, Bangkok, and Nairobi. SITREPS are posted for public access, and individuals and organizations can subscribe to receive SITREPS. The website and supporting structure is designed to filter data from trusted sources across the response, and provides a wide variety information in standardized data. These website reports include news articles, public reports, press releases, appeals, policy documents, analysis, and maps related to humanitarian emergencies worldwide. Another similar product is the humanitarian dashboard (https://assessments.humanitarianresponse.info/humanitarian-dashboards), a web-based tool to allow the individual clusters within a response team to monitor the status of the implementation of the disaster plan. It is designed to provide an overview of gaps in the response and to allow the individual clusters to better align and coordinate their efforts.

Both of these structures (the incident command system and the U.N. cluster approach) use space technologies, imagery, data, and telecommunications as a major aspect of their work. Likewise, the many NGOs and volunteer relief organizations also use a variety of space technologies and information systems as well, independently or as a part of the overall response. This is particularly true for the Red Cross/Red Crescent, but can also include CARE International, Oxfam, and others.

You can see through this brief explanation of the organization and structure of groups that respond to a major disaster are complex. It is very difficult, under the best of circumstances, to properly coordinate actions and share resources and information. In a real disaster events move quickly. Lives and property are at stake, and critical infrastructure can be damaged or destroyed. In this environment information sharing can become an overwhelming challenge. Space technologies can play a major role in all aspects of these activities, but only if the emergency responders across all organizations are properly trained and able to utilize these resources effectively.

Chapter 4
Space Systems for Disaster Management

Our access to space and the proliferation of satellite space systems lend important capabilities to global disaster rescue efforts at all phases of the cycle. Satellites provide vital severe weather warnings, telecommunications, mapping, positioning, navigation, and imagery on a global basis.

In the preparation cycle, various remote sensing data are used for weather monitoring. Further, remote sensing imagery are now being consistently used in planning for disaster recovery. GIS datasets are analyzed, and satellite telecommunications systems are routinely employed around the globe for data and communications in all four phases of the disaster management cycle.

Once an event occurs, satellite telecommunications are often the only reliable means of communicating, as ground and cell systems are often destroyed. High resolution imagery provides damage assessment, GPS provides mapping and navigation, and GIS is used throughout the initial response. These continue in the recovery phase as well. In the mitigation stages, many types of application satellites also play key roles in minimizing the effects of future events. These include fixed and mobile satellite telecommunications systems, weather satellites, positioning, navigation and timing satellites, plus a variety of satellite imaging satellites.

Current space remotes sensing satellite include spacecraft that range from geostationary weather systems that are constantly on guard for severe weather, moderate resolutions systems that map Earth daily, and very high spatial resolution satellites in lower orbits that can acquire images down to 35 cm (or objects about one foot in size) for detailed mapping of resources and infrastructure. RADAR satellites can acquire detailed imagery at night and through clouds and rain, providing vital quick response in weather-related incidents.

Satellites also provide in-situ monitoring and telemetry systems for data such as ocean buoy networks and flood monitoring stations. Geographic Information Systems (GIS) power high-end computer flood modeling, and much more. Precision Positioning, Navigation and Timing (PNT) satellites such as the U.S. GPS, Glonass, Compass, and Beidou can provide real-time positioning and navigation for responders, track vehicles and shipments of relief supplies, and can be used to map damage in the field as soon as responders arrive on the scene. Finally, GIS integrates all of these various sources of information into a single, integrated system that drives the situational awareness and command and control aspect of the response. All play important roles individually, but it is when they are integrated that we derive the maximum benefit.

Disasters come in many shapes and types. Some are without warning, such as earthquakes, but others, such as cyclones and hurricanes, can be tracked using geostationary weather satellites, providing vital warning and evacuation time. Once an incident occurs, each of these technologies, alone and also integrated together, provide important capabilities. Satellite telecommunications are often the most important immediate capability, when a disaster incident renders the existing land and cell systems unusable. Therefore we will address satellite communications systems first. This is not only because they are often needed immediately, but because this involves the most diverse technology with the most service options, and the most difficult financial and contractual arrangements.

If there can be a standing order and contractual arrangements for these services, then emergency link-ups by satellite can be expedited. Moderate and high-resolution satellite imagery provides important 'before and after' change detection that often gives the first overall view of the scope and coverage of the incident. GPS systems are used to map the impact and to track response vehicles and assets. All of this information and more can be integrated into a GIS, which is the situational awareness and command and control environment. There are several new and evolving technologies such as social media (not specifically space assets, but used in association with them) and community remote sensing that can provide additional, vital information.

Satellite Telecommunications

First and foremost, satellite telecommunications networks provide important voice, data and Internet access in areas where existing services are knocked out or did not exist. Cell, voice, and Internet services are often destroyed or significantly reduced in the initial stages of a disaster, and satellite telecommunications provide the important, immediate access to voice, data, and Internet in the early stages of the response. Frequency management is an important aspect of this, and it is vital that responding organizations provide their own capabilities and be fully trained in their use.

These communications facilities for emergency response might be hand-held mobile satellite phones or laptop units that just require activation and occasional recharge from an emergency generator. Broader band communications might be provided by fixed satellite service (FSS) that offer rapid setup and deployment of very small aperture terminal (VSAT) satellite dishes. Sometimes these have to be assembled from suitcase-sized packaging units. Some emergency response VSAT are inflatable units that need a generator and pump to get them into working order.

Geostationary weather satellite data are utilized to provide long-range warning of severe weather incidents, as well as daily weather information for the affected area. Moderate resolution imagery such as MODIS and Landsat provides quick 'before and after' comparative data to create general maps (as may be required) to show impacted or flooded areas, and provides a baseline for initial damage assessment. This is generally geared towards determining what specific types of resources and responders are needed. High-resolution optical imagery is now routinely used for initial damage assessment, locating appropriate shelter and helicopter landing areas, and detailed mapping of the disaster situation. It is also used for more precise mapping of the situation before the incident and then comparing this with imagery taken after, in order to conduct damage assessment.

New, high resolution RADAR imagery provides important mapping and impact analysis, especially in cloud-covered areas and for infrastructure that shows up particularly well on radar imaging, such as roads and rail systems. Global positioning, navigation and timing (PNT) data such as the U.S. GPS system Glonass, and other new systems are now an integral part of the response, being used for damage assessment, mapping affected infrastructure, and routing, tracking and locating response vehicles.

GIS really is the glue that brings all the data together and is the environment that collects imagery, GPS, and field mapping information for a variety of purposes, including providing daily situational maps, supporting press briefings and hundreds of additional other uses by the emergency responders. We will now review each of these important capabilities in more detail to see what each brings to the disaster response.

Satellite Telecommunications Systems Capabilities

Disaster planners and responders have to be prepared for the destruction of communications infrastructure in the incident of a disaster. The use of satellite communications is quite often a key part of the COOP. It is clear that the disaster response depends on good communications, and disaster managers can only make decisions based on what they know. Lack of communications can result in poor response decisions and unneeded loss of life and property. All EM agencies are aware of this, and there is a need for independent communications, training, and capabilities. Responders also need interoperability and compatibility of communications networks, and there are several tragic examples of different responders not being able to communicate using their individual radio systems.

There are several common problems, including the actual destruction of the satellite, ground, and cell infrastructure and electrical systems. Of those systems that survive they are often overwhelmed after an incident, when they are saturated by people seeking assistance or trying to contact family members. This often blocks access to the system at these critical times by responders, so many EM organizations provide their key members with satellite telephones to ensure communications. Very small aperture terminals that access fixed satellite networks can also be deployed from small, suitcase-sized systems. There are even inflatable VSAT

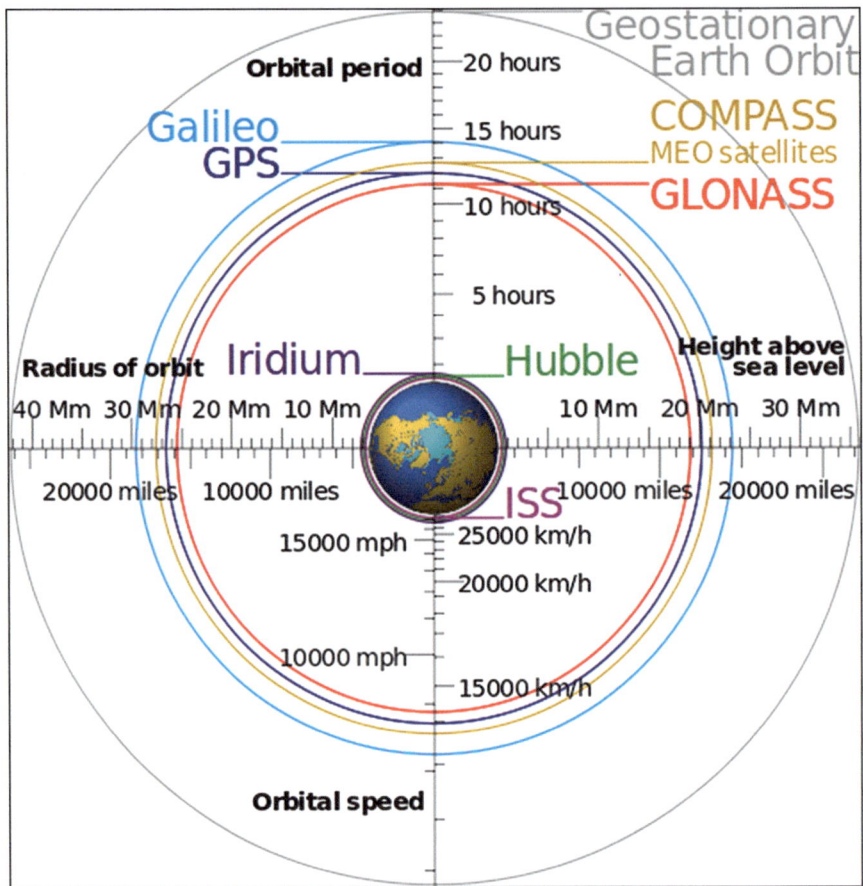

Fig. 4.1 Satellite orbits and velocities for different types of satellites (Graphic courtesy of NASA)

antennas that can also quickly deployed. In short, mobile and fixed satellite systems are key to providing immediate emergency responder communications.

Satellite telecommunications systems are a very large commercial business, with over US$185 billion in launch, manufacturing, ground stations, and services each year. Currently satellite services are still growing at a significant rate. The satellite telecommunications service sector globally is over a US$125 billion business. Telecommunications services are generally provided within three types of systems, defined by their orbits. These are LEO (low-Earth orbits), MEO (medium-Earth orbits) and GEO (geostationary Earth orbits). GEO systems do not require continuous tracking and pointing to satellites, as satellites in this orbit appear to constantly hover above the same point on the equatorial plane.

Figure 4.1 shows the LEO, MEO, and GEO orbits, with their orbital periods, radius, orbital speed, and height above sea level. The Iridium satellite is shown in

LEO orbit with a very rapid orbital speed. There are various positioning, navigation, and timing (PNT) satellites also shown in MEO orbit. Most of all telecommunications are positioned in the special GEO orbit way out almost a tenth of the way to the Moon. Nevertheless there are telecommunications satellites in LEO and MEO altitudes as well.

GEO: The Geostationary Telecommunications Satellites

The geostationary orbit is circular and located 35,786 km precisely above the equatorial plane. Satellites in this orbit rotate Earth at the same speed as the planet revolves, and follow the rotational direction of the planet. This provides a unique location for providing global telecommunications, meteorological observation, and remote sensing capability because the orbital period is the same as Earth's rotation, so that a satellite in this particular orbit appears to remain in the same location hovering over the same spot on the equator. This allows an antenna on Earth to point to a fixed location in the sky without actively tracking the satellite.

Satellites in GEO can provide voice, video, and broadband data to a broad area of the globe, i.e., nearly a third of the planet. This means that three equally spaced satellites can broadcast to the entire planet. This is sometimes referred to as the Clarke orbit, in honor of Arthur C. Clarke, whose 1945 paper first proposed the use of this location for telecommunications satellites.

Figure 4.2 shows the ever more numerous satellites in the GEO orbit. These are distributed around the orbit, but also clustered over main population centers of the world such as western Asia, western Europe, and the Americas. There are over 20,000 individual satellite television channels in the GEO orbit on either FSS or BSS spacecraft alone. Some communications satellites include as many as 100 transponders, although 24 to 48 transponders are more typical. A transponder is typically 36 to 72 MHz in size and can support the transmission of many hundreds of voice channels or over a dozen television channels using digital transmission technology. These telecommunications satellites operate in a number of frequency bands in the VHF and UHF range for mobile transmission as well the SHF and EHF bands for fixed and broadcast satellite services. The frequencies used for satellite communications can range as low as 137 MHz (in the VHF band) up to 30 GHz (in the Ka band).

The number of satellites in the GEO orbit, especially in key locations above the United States, Europe, and Asia, have led to orbital congestion, particularly in the radio frequency bands assigned to satellite communication. Orbital congestion for the FSS transmissions in the GEO satellite plane is most severe in the C-bands (6/4 GHz) and Ku-bands (14/12 GHz). Even the demand for GEO services for Ka-band (30/20 GHz) is rapidly growing. Thus future congestion in this band will occur in coming years as well. The congestion is more in terms of radio interference than it is of potential orbital collision. Satellites that are separated by 1° on the GEO plane are

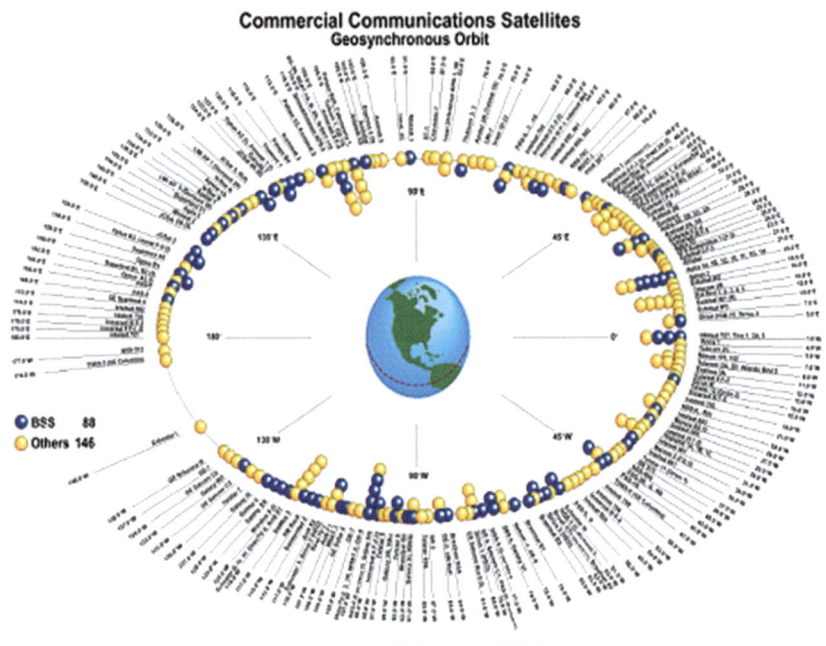

Fig. 4.2 Orbital congestion in the GEO orbital plane (Graphic courtesy of NASA)

actually about 710 km, or 435 miles, apart. Satellites in GEO that operate in the same frequency band are separated by at least 2° in order to avoid signal interference.

Broadcast satellite systems only require a few satellites to provide service since they are one way transmission systems. Digital modulation, multiplexing and compression techniques are used by almost all types of satellite communications systems in order to be more efficient and to use less spectra to accomplish their tasks.

There are three primary services provided by GEO satellites: fixed satellite services (FSS), mobile satellite services (MSS), and broadcast satellite services (BSS). FSS service is provided between and among fixed antennas with a clear line of sight to the satellite. These typically operate with a 3 to 6 dB of link margin. This means an extra margin of power that is two to four times that needed to complete the link. The margin is provided in case of interference or attenuation due to rain or other precipitation (called rain fade). Mobile MSS systems have to overcome all of these difficulties and more. They must transmit to and receive from a small, low-powered mobile transceiver. They must have much higher power margins to ensure continuity of service since obstacles can get in the way of the satellite transmission, obstacles such as trees, buildings, signs, and others that a moving vehicle encounters.

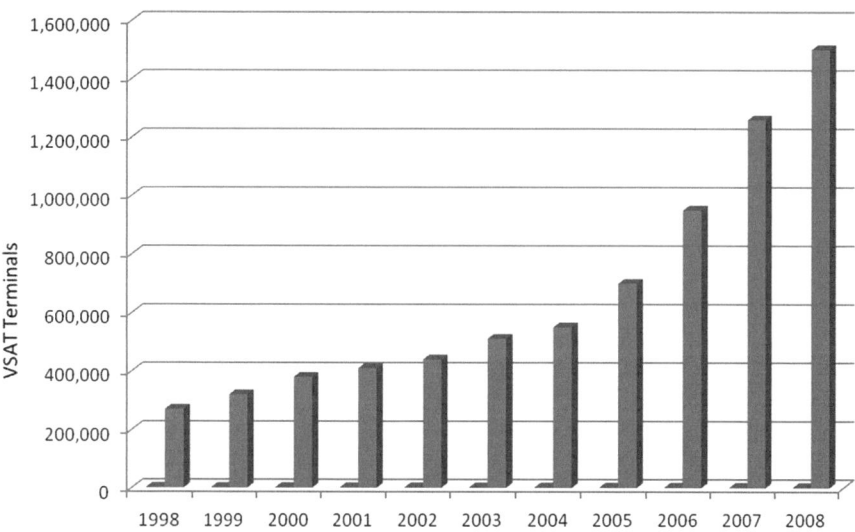

Fig. 4.3 Growth of VSAT antenna use continues unabated (Graphic courtesy of Joseph N. Pelton)

BSS systems have to provide sufficient downlink power so that the user terminal can be extremely small and very low cost (such as small 'pizza dish' satellite TV antennas). Also these systems have to obtain access to programming and deal with issues such as copyright protection, pirated services, etc. As FSS systems have become more and more distributed and have begun to provide services to end users, they have found they have to deal with these far more complex legal and policy problems as well. The growth of Internet protocol (IP) based digital services and the increasingly capable VSAT antennas – that are very cost effective and easy to use – have contributed most to the FSS growth over the past few years.

Figure 4.3 shows the growth in VSAT terminals around the world. The adoption of new Internet Protocol over Satellite (IPoS) standards has accelerated this trend of business and small office and home office (SOHO) use, with VSAT network for a growing range of uses.

Fixed Services (FSS)

This provides broadband services to fixed receiving stations around the world. Services include trunk cable TV, Internet, data, voice, and other broadband services. Fixed satellite services are the easiest to establish and maintain, because the antennas (that are transmitted to and receiving signals from the satellite) can be fixed in placed so as to have a clear line of sight to the satellite. They can have a constant look angle to the satellite and thus minimal steering (assuming a GEO satellite). These FSS ground antennas can often be positioned in protected areas so as to minimize radio frequency (rf) interference. Since they can be hooked up to a constant power supply, they do not have to depend on battery power, except possibly to restore service in case of a power outage. For the last 5 years Internet/IP-related services and especially digital video broadcast with return channel service (DVB-RCS) and other similar standards, including the cable television-based standard of DOCSIS (Digital over Cable System Interface Standard) have been a primary growth area for FSS satellites. As noted earlier, these systems operate in the C band (6/4 GHz), Ku band (14/12 GHz) and, more recently, Ka band (30/20 GHz). In addition, defense-related fixed services are primarily in the UHF band (320/240 MHz, X band (8/7 GHz) and Ka band 30/20 GHz). The communications satellites that operate in the higher frequencies (especially above 20 GHz) can be degraded by heavy rainfall and other atmospheric conditions. Therefore greater power margins are usually provided for Ka band satellites.

The FSS sector of the satellite industry has had more limited revenue growth for the past few years, unlike the BSS industry that has exhibited very rapid growth and is now the largest generator of satellite revenues. The global economic downturn dating from 2004, the overbuild of capacity by fiber optic networks that depressed telecommunications pricing, and the much greater capacity available via new, so-called "high throughput satellites" (HTS) have all served to reduce the profitability of the FSS industry. Over the longer term the new high throughput satellites are expected to improve satellite revenues and their competitiveness with regard to fiber and coaxial cable-based terrestrial networks.

Newly emerging business models based on high throughput satellites and more efficient use of satellite capacity, including new digital compression technologies, is setting the stage for a new era of growth over the next 5 years, which will ultimately increase FSS revenues and profitability.

Mobile Satellite Services (MSS)

Mobile satellite services are, as the name implies, provided to receivers on the ground that are located on moving vehicles such as trucks, aircraft, or ships at sea. The user unit might even be a small hand held or laptop unit that an individual might use. These services are offered in the lower VHF and, UHF bands (137 MHz, 400/432 MHz, 1.6/1.5 GHz, 2.0 GHz, 2.5/2.6 GHz bands) because mobile terminals may be shadowed or partially blocked due to their mobile nature. These mobile satellite networks benefit if they can use lower frequencies with larger wavelengths that are more tolerant of not always having a clear line of sight to the satellite. In short, the larger wavelengths in the lower frequencies can "see" around obstacles in the way that higher frequencies with very short wavelengths cannot.

As previously noted these mobile satellite systems also need higher link margins (10 to 22 dB) because of blockage by foliage, power line poles, buildings, hillsides, etc. They also need more power (power flux density) to be received, especially by small portable or mobile user terminals. Although mobile satellite services represent the smallest business market in satellite telecommunications these systems are currently demonstrating the largest percentage growth among the three primary satellite services.

Mobile satellite services can be provided from LEO and MEO constellations or from GEO. Since the mobile user on aircraft, boats, or on the ground (walking or in vehicles) are moving the antennas used to access the satellite must be able to track the satellite or have the ability to receive the satellite signals at all angles. These mobile terminals for users, because they must be small and typically receive signals on the move and at different angles, are not able to provide data rate signals in as broad a band as those that can be maintained by FSS or BSS systems. (One can review other books in the Springer Press-International Space University series such as *Satellite Communications* by Joseph N. Pelton to learn more about the nature of the ground segment and the space segment for fixed, mobile and broadcast satellite networks.)

Broadcast Satellite Services (BSS)

This is the service that provides direct broadcast of broadband Internet, television, and radio as a one-way service. This signal can go to businesses, to homes, and, increasingly, even to laptops and mobile handheld devices. Broadcast satellite services (BSS) are offered around the 18/12 GHz bands allocated by the ITU. DirecTV and Dish in the United States, BskyB and Eutelsat Hotbird in Europe, NHK's Broadcasting Satellites in Japan, and Nimiq in Canada are just some of the many DBS services around the world. The size of user terminals today for BSS are in the 0.8 to 0.35 m range.

What is called direct-to-the-home (DTH) services are sometimes offered on FSS satellites that have higher power. These can offer television and radio service to a small receiving terminal at the home even though it is technically not a BSS satellite operating in the BSS frequencies. Hughes Galaxy, SES Astra in Europe, and PanAmSat have operated DTH service in FSS frequency bands. BSS and FSS satellites are both used to provide digital video broadcast (DVB) services to distribute high-speed data.

There are a number of terms used to describe 'broadcast' services. There is the ITU term that is broadcast satellite services (BSS), but this same service is also sometimes called direct broadcast satellite (DBS) service. Then the term direct-to-the-home service is used to describe quasi-DBS services that are offered using FSS satellites with very high gain. This DTH service is still prevalent in Europe (i.e., the Astra satellite services), but this is now less so in the United States and Japan. In the United States this is a service sometimes called backyard dish service because the terminals must be larger to obtain good quality reception over the larger coverage areas of the United States. There is also an issue with DTH as to payment for the service. It is illegal to buy de-scramblers and not pay the distributors of the service and the premium channel providers for this service. This is, however, sometimes hard to enforce. Therefore the providers have moved to more sophisticated digital encryptions that makes unauthorized reception more difficult.

Although this part of the satellite industry is hugely profitable and is global in scope, it is less relevant to disaster response and communications because it is a one-way system that distributes information rather than operates as an interactive system. There are some BSS systems that do support what is called a digital video broadcast with return channel services (DVB-RCS) or another standard first developed for cable television called DOCSIS. When a BSS system operates in this mode it essentially becomes a fixed satellite service.

The FSS sector of the satellite industry has remained somewhat static for the past few years, unlike the BSS industry that has exhibited very rapid growth and is now the largest generator of satellite revenues. Although the FSS industry grew to over $11 billion in revenues by 2004 this sector has been adversely affected in the past 4 years by the general downturn in the technology and telecommunications sectors and the overbuild of capacity by fiber optic networks on a global basis. Today, a decade later, FSS industry revenues have managed to grow to $15 billion per annum. New business models, high throughput satellites that allow for more efficient use of satellite capacity, as well as the latest digital compression technologies is setting the stage for a new era of growth over the next 5 years.

Figure 4.4 below shows some of the many international communications satellite systems currently available that can be used for emergency and disaster response communications. It should be particularly noted that there are now domestic mobile and FSS satellite systems that have been deployed, especially in the United States, which can be and indeed have been used in emergencies. These systems include Via Sat and the SkyTerra system.

The term link margin, measured in decibels (dB), is the difference between the received power required at the receiver and the actual power received. A 10 dB link margin means that the system has ten times more power than needed to complete the satellite link but could still (barely) operate with 10 dB less signal power.

GEO Telecommunications Systems

There are two primary, widely available solutions for disaster communications via satellites, as well as new systems that are becoming available. The two currently available technologies are VSAT (very small aperture terminal) systems, which operate on a variety of satellite systems, including Intelsat, SES, Telesat, and others, and the BGAN and Global Xpress system from INMARSAT.

VSAT (very small aperture terminal) operates on a variety of satellite systems, including Intelsat, SES, and Telesat, among others. VSAT systems use 'pizza dishes' ranging from under 1 m up to 2.5 m and provide between 64 k to 72 megabits per second, providing broadband for local wireless IP systems. VSAT systems can support large field operations with dozens of concurrent users. Entry-level equipment can cost around US$3,000 and use costs can range anywhere from US$200 to

Important International Satellite Networks	Key Operational Features
Eutelsat	This is a regional GEO system that is expanding to become a global system. It provides service to VSATs and microterminals.
Inmarsat (MSS)	Global coverage – voice, data, whiteboard, video, to laptop or handheld units. It supports mobile communications or can work with fixed site microterminals. It uses Inmarsat 4 Geo satellites to support very small handheld and laptop units to speeds up around 500 kilobits/second (about 10 dB link margin). By 2014 the Ka-band Global Xpress Satellites were being deployed to offer even higher data rates.
Globalstar (MSS)	Wide global coverage – voice and data, limited video to handheld units from a constellation of low Earth orbit at lower data rates, but the new generation of GEO satellites will support higher rates (8 to 10 dB link margin).
Iridium (MSS)	Full global coverage – voice and data to handheld unit at lower data rates, but next generation will support higher data rates (10 to up to a 16 dB link margin). Unique feature is service to the entire globe including the higher latitudes near the poles.
Thuraya (MSS)	Coverage of parts of Europe, Africa, Mideast, Asia, and Indian Ocean from very large aperture GEO satellites. Voice, data, video, to laptop and handheld unit from GEO satellites (6 to 10 dB link margin).
Intelsat (FSS)	Global coverage – voice, data, video to VSATs and micro-terminal (typically 3 to 6 dB link margin, but higher margins for the Ka-band service) GEO satellites at many locations and in a variety of frequency bands (C, Ku and Ka bands).
O3b	This is a new constellation of medium-Earth orbit satellites that is optimized for the equatorial region and IP-based services. Just being deployed in partnership with SES, Google, Liberty Media, etc.
IPStar (FSS)	Global coverage – voice, data, video to micro-terminal (3 to 6 dB link margin). Provides service to VSATs and micro-terminals via GEO satellites.
SES Global	A large variety of GEO satellites provide FSS and DTH and BSS on a global basis. It also provides service to VSATs and micro-terminals

Fig. 4.4 Key examples of major international satellite networks (Graphic prepared by the author)

Fig. 4.5 A portable VSAT system in use by a U.S. military unit (http://www.dvidshub.net/image/50091/vsat-helps-eod-save-lives-time#.Upo_dmR4ajd)

$20,000 per month, depending on bandwidth. The system can be used worldwide, but require a contract with individual satellite providers, depending on location. The systems are not for mobile applications, requiring a fixed dish pointing to the correct satellite, but it can support a wireless IP network for mobile users. Systems use GPS to properly locate the nearest satellite and can operate from a vehicle battery, generator, or power outlet, and can be set up in less than 15 min (Fig. 4.5).

Broadband global area networks (BGAN) from INMARSAT provide excellent mobile broadband for disaster situations, costing from between US$3,000 to $15,000 per system and consisting of a small, flat panel terminal the size of a laptop computer. These networks provide high-speed data transfer, with speeds approaching 500 kb/s, with worldwide coverage and guaranteed data rates. It supports both IP (Internet Protocol) and circuit switched (traditional telephone) applications using a small terminal the size of a laptop that can run from a battery and is designed more for individual users or small groups with small bandwidth requirements than VSATs.

There are a wide range of BGAN terminals available that all provide Internet and telephone, including three BGAN satellites that provide global coverage, except at the extreme polar areas, and any terminal will work with any of the three INMARSAT satellites, spaced evenly around the world. Data cost is significant, ranging from US$3 to 7 per Mb uplink. The same INMARSAT satellites provide the I-Sat mobile phone service (Figs. 4.6 and 4.7).

The Globalstar mobile satellite service has converted from being a low-Earth orbit constellation to its network design. This newest configuration uses three geostationary satellites equally spaced around the equator to provide near global coverage of Earth, only excluding the extreme polar regions (Fig. 4.8).

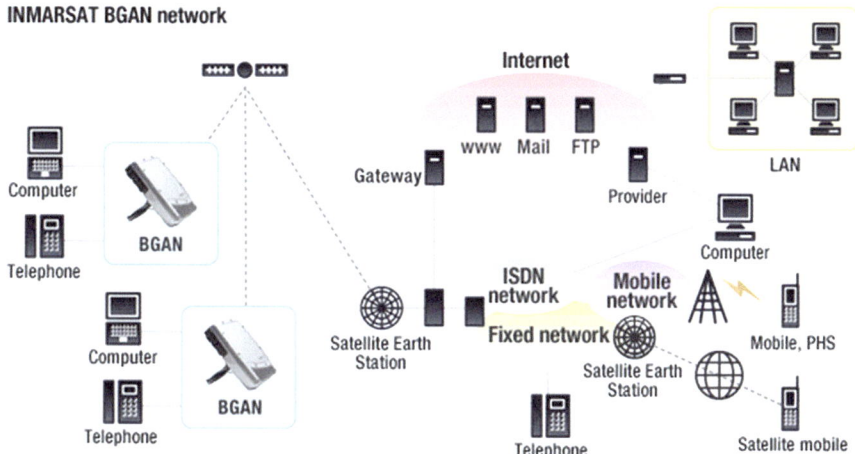

Fig. 4.6 A typical schematic diagram of an Inmarsat BGAN network

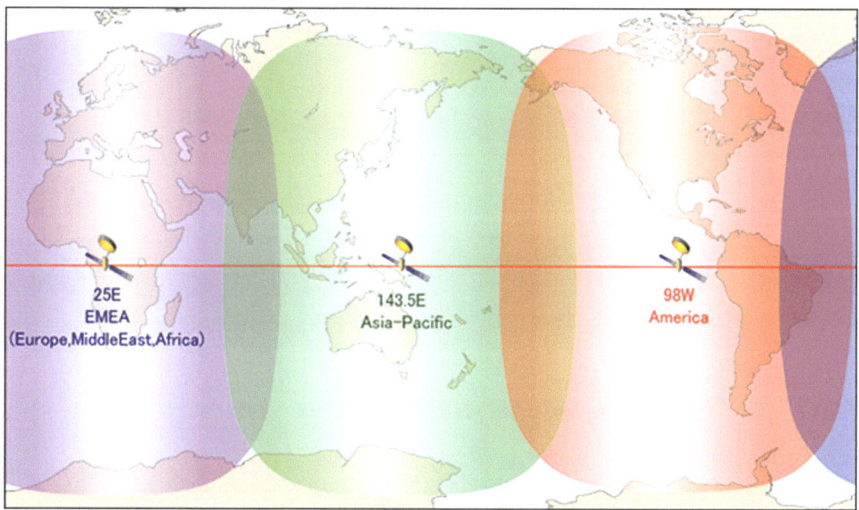

Fig. 4.7 The latest generation of Globalstar satellites (Image courtesy of Globalstar)

MEO Satellite Telecommunications Systems

MEO orbits are primarily the domain of GPS and similar navigation satellite constellations such as the Russian Glonass, European Galileo, and others. It lies between LEO and GEO, and many of the MEO navigation satellites are around 20,200 km with a 12-h orbital period, about one half the period of the GEO satellites.

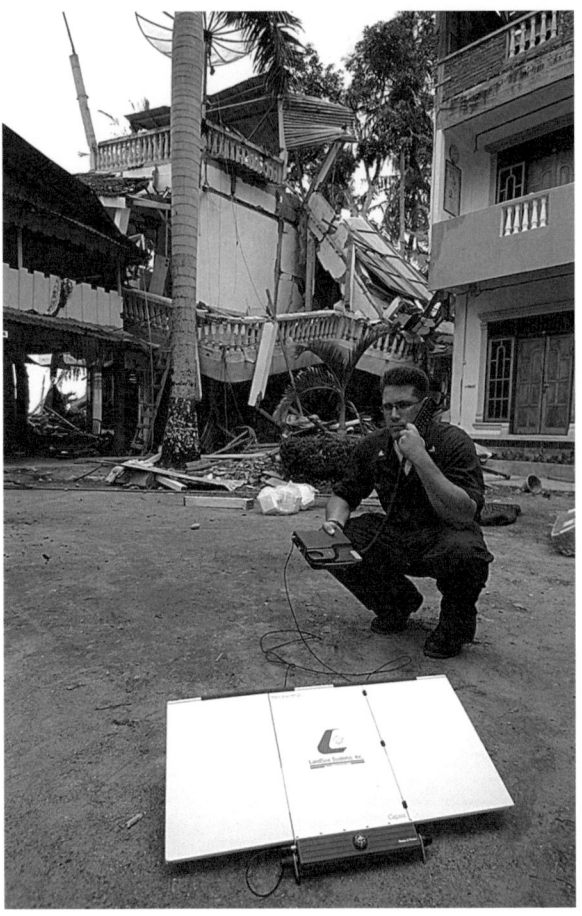

Fig. 4.8 A U.S. Navy sailor using an Inmarsat telephone in Indonesia in a disaster (http://en.wikipedia.org/wiki/File:Satellite_phone.jpg) (Image courtesy U.S. Navy)

LEO Satellite Telecommunications Systems

Low-Earth orbits are typically more than 500 km up and less than 1,000 km. This area traditionally has not been used for telecommunications, but lately this has changed, and we now have several LEO telecom systems.

Iridium

The Iridium system is a low-Earth orbit constellation of 66 satellites (plus in-orbit spares) that are in a 780-km orbit inclined 86.4°. This large number of satellites and their low orbits permit lower power, smaller antennas on the mobile phones, and less delay in the signal as well. The satellites can link with other satellites in their

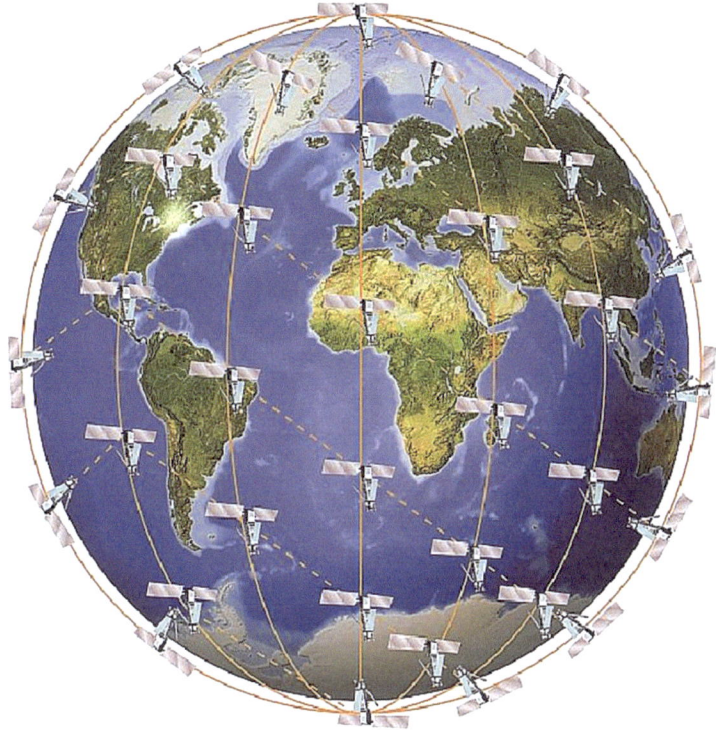

Fig. 4.9 The orbital coverage of the 66 Iridium NEXT constellation (Image courtesy of Iridium Satellites)

own and adjacent planes to 'pass along' calls. This is required as the LEO satellites quickly pass out of range of any location on Earth. Iridium provides a truly global constellation for disaster response use and provides equal capabilities anywhere on the planet. The phones can be used while mobile, but, like all satellite systems, must have a direct line-of-sight with the satellites.

In 2015 the next generation Iridium-Next system will be launched. This new network will also consist of 66 satellites, and will provide global voice and data coverage, but with expanded capacity, higher data rates, and improved link margins (Figs. 4.9 and 4.10).

Globalstar

Globalstar is another LEO telecom system that has a constellation of 48 satellites inclined at 52°, and which has been operational since 1998. It is a simple 'bent pipe' system, with no satellite-to-satellite linking as in Iridium, and provides a different service, that of mobile and fixed services, and also has a handheld satellite messaging

Fig. 4.10 U.S. Air Force crew using an Iridium phone in Antarctica (http://upload.wikimedia.org/wikipedia/commons/b/b4/USAF_iridium_phone_use.jpg)

and tracking system called SPOT Satellite Messenger. It is widely used in disaster management, but requires a local gateway station, as there is no intersatellite link. The latest generation of Globalstar satellites, however, are deployed in GEO and thus the low-Earth orbit is being phased out.

Orbcom

Orbcom is another LEO system that provides machine-to-machine (M2M) data. This constellation relies on a constellation of 29 satellites, orbiting at 775 km, using small, 42-kg satellites that provide 4,800 bits/s downlink and 2,400 bits/s uplink operating in the VHF band. Orbcom provides satellite data services, not voice, and includes email and messaging services, which are very useful in disaster situations. Orbcom satellites are often used for ground- and sea-based fleet management and thus user units often combine Orbcom and space navigation capabilities.

Satellite Telecommunications Disaster Applications and Issues

In hurricanes Katrina and Rita in 2005, satellite technologies provided vital communications capabilities. A full three weeks after Katrina made landfall, only 60 % of the area's cell service was operating, over 1,000 cell towers were still down, a

Fig. 4.11 Red Cross VSAT facility in Typhoon Haiyan (Image courtesy the American Red Cross)

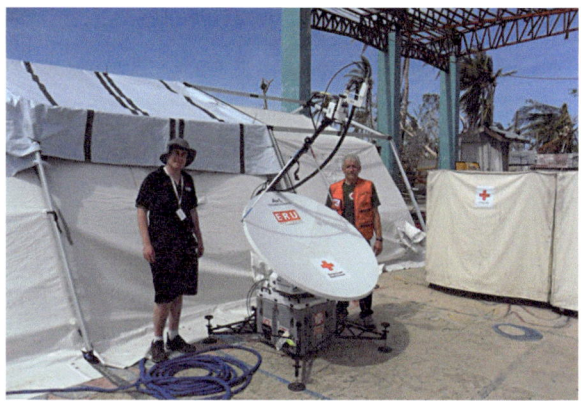

full 30 % of all media broadcasts were not functioning. Over two million telephone calls were failing each day, and over three million telephone customers were still without service. But satellite services were fully functional, according to FCC inquiries after the fact. The Red Cross deployed all nine of their emergency communications RVs (ECRVs) (See Fig. 4.11). In this one emergency event, over 20,000 Globalstar, Iridium, and Mobile Satellite Ventures phones were deployed. Iridium traffic in the gulf area increased over 3,000 %.

VSATs, or very small aperture telecommunications systems, helped to provide quick Internet, fax, web-based applications, telephone and videoconferencing. During Hurricane Katrina, the American Red Cross established a 50+ node VSAT network that provided over 5,000 voice over IP (VOIP) telephone calls per week, videoconferencing, Internet, and web-based applications and email.

After typhoon Haiyan in 2013, Intelsat provided a Ku-band beam on its Intelsat 8 satellite for Philippine relief effort. This included bandwidth for the American Red Cross International Emergency Response unit in the Philippines and also to the non-profit Global Disaster Immediate Response Team, which provided inflatable satellite terminals that linked to the Intelsat 19 satellite.

It is vital that responders be able to identify available inventories of telecom and IT capabilities and capacities. Frequency allocation and alternate plans for capacity loss is also important, as are plans for pre-positioning equipment whenever possible. Identifying partners who have the ability to repair existing infrastructure and deliver additional capacity is vital. First responders and others involved in the response effort must be trained in the use of all equipment, and realistic exercises using reduced capacity and the introduction of satellite capacity is important. There is one unfortunate example in hurricane Katrina in Louisiana, where large numbers of satellite telephones were available and needed, but were never used because responders were not able to operate the equipment due to lack of training.

An additional area of significant benefit is that of tele-health and tele-medicine, in disaster situations, where medical services such as radiology can be provided remotely from facilities well outside the impacted area.

Fig. 4.12 Red Cross emergency communications response vehicles (ECRV) (Graphic courtesy of the Red Cross)

The American Red Cross

The American Red Cross operates nine American Red Cross Emergency Communication Response Vehicles (ECRV), as shown, that were donated by Ford Motor Company, with 16 HF, VHF and UHF antennas, a VSAT antenna and 52′ pop-up mast, and 10 satellite telephones. These vehicles were heavily used beginning in 2002. Today, however, they are being phased out for updated systems that can be used with any Red Cross or rented vehicles (Fig. 4.12).

Satellite Telecommunications Emerging Technologies

Large aperture multi-beam antennas allow the creation of many highly concentrated beams that not only created concentrated beams of higher power density but also allowed spatial separation so that radio frequencies can be more efficiently re-used.

Satellite power has increased not only by the deployment of larger solar cell arrays on today's satellites but also by the use of higher efficiency solar cells in terms of converting a higher percentage of received solar power to electrical power. Also improved batteries (such as lithium ion) allow longer life and support operation during solar eclipses. High gain, multi-beam antennas only work effectively if they can be continuously and accurately pointed toward Earth's surface. Therefore today's satellites have much greater pointing accuracy as well. All of these innovations allow user terminals to shrink in size, cost, and portability.

One of the key challenges is to find ways to meet broadband telecommunications needs within available spectrum. One approach may be to use satellites to extend wireless networks to meet the needs of rural and remote areas where population densities do not make conventional terrestrial systems economic. Cellular backhaul, WiFi extensions, and even rural wireless hubs can be supported by today's fixed satellite systems (FSS). A combination of FSS and rural wireless systems are key to service in areas such as Africa.

New systems such as O3b that have been optimized for IP-based systems are well suited to such applications.

South America and Asia

One of the true keys to satellite success in the broadband IP world is the development and implementation of standards that allow efficient transmission of Internet/IP services in spite of latency and other difficulties associated with space communications. These standards adopted in November 2003 and January 2005 are now being widely accepted. IP over Satellite (IPoS) accelerators to move IP traffic efficiently over satellites are being applied to most VSAT systems now in operation.

Unmanned Aerial Vehicles (UAVs) or High Altitude Platform Systems

Unmanned Aerial Vehicles (UAVs), or in the parlance of the International Telecommunication Union (ITU), high altitude platform systems (HAPS), can provide excellent telecommunications, Internet, and remote sensing capabilities in disaster areas by loitering at very high altitudes such as 21 km (about 14 miles) over an affected area for extended periods of time, ranging from days to weeks. Such systems could replace destroyed cell tower, microwave relay and telephone service over a wide area. Several prototypes include the hydrogen-powered Boeing Phantom Eye and several solar-electric vehicles are in development, but none are in service yet. These could provide local telecom and remote sensing services in areas that have been devastated, but there remain several technical, institutional, and legal hurdles before so-called Protozone-based high altitude platform systems (HAPS) can become a regular component of the disaster response.

Chapter 5
Space Remote Sensing Fundamentals and Disaster Applications

Satellites provide a superb vantage point for viewing our world. This is done through the technology of remote sensing, or collecting information at a distance. Remote sensing is done by the gathering and analysis of electromagnetic (EM) energy, and by analyzing the interaction of such energy with a target such as the surface of Earth. This can be done from a variety of orbits and can provide many different types of useful information in a disaster.

The electromagnetic spectrum is a continuous spectrum of energy, ranging from very high energy gamma rays and X-rays to very low energy radio waves. All EM is exactly the same, and only differs in the wavelength and amount of energy contained. EM can propagate through a vacuum, which makes remote sensing from satellites in space possible. The energy we see with our eyes lies in the visible spectrum, which is a very tiny portion of the overall spectrum. We have created many different detectors and sensors that allow us to collect and analyze energy from throughout the spectrum, allowing us to 'see' far beyond our ability to see without eyes. This multi-spectral remote sensing provides powerful support to disaster planning and relief operations.

Optical, or passive remote sensing, satellites are very complex, and collect data that we see as images by acquiring energy that has been reflected off the surface of Earth and passing this energy through a telescope and into a detector. The detector, using various filter and prisms, separates the energy into specific spectral bands. Such systems collect data from the ultraviolet (UV) through the visible area, and into the infrared (IR) and thermal infrared (TIR) regions. We record the intensity of these data and transmit them to Earth for further analysis and processing. By combining three bands of data we can create color images that can represent 'true' colors as we would see from space, or any combination of bands beyond our eye's ability to see, which are called false color composites. Imaging in the thermal band allows us the 'see' heat, while microwave remote sensing allows us to acquire imagery through clouds and at night.

Each satellite sensor has four parameters of resolution, which define both the capabilities and uses of that sensor. **Spatial resolution** is both the smallest area

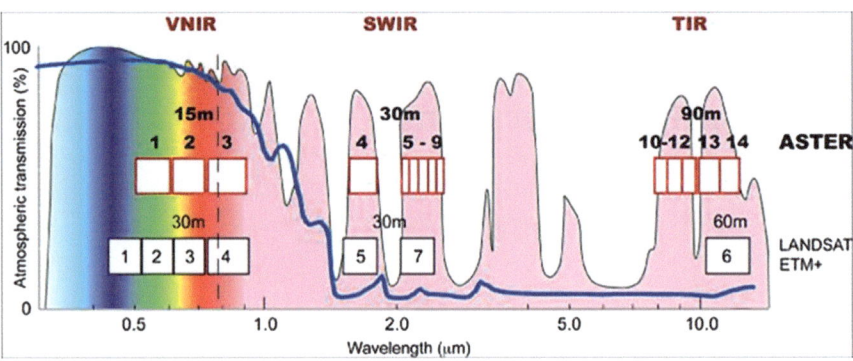

Fig. 5.1 Landsat and ASTER spectral bands (Image courtesy of NASA)

where data are collected, and also the area covered in one pass, or swath width. Geostationary weather satellites may only have a spatial resolution of 1 to 5 km, where ultra-high 'spy' satellites have resolution of less than 1 m. **Spectral resolution** is the portion of the spectrum where data are collected. **Radiometric resolution** defines how finely the data can be quantified and is measured in bits, where 8-bit data are quantized in a range of values from 0 (no energy entered the sensor) to 255 (where the sensor was overloaded). Current sensors range from 8- to 12-bit radiometric resolution. **Temporal resolution** represents how often the satellite acquires data over the same location on Earth. GEO weather satellites can image Earth every 5 to 15 min, while some high resolution satellites can take weeks to revisit the same location. These four resolution parameters define the capabilities and uses of the data. All remote sensing data are processed into a raster format, where each pixel represents a single data measurement on the ground. These raster images can then be processed, analyzed, and entered into a geographic information system for integration with other data.

Figure 5.1 shows the regions of the spectrum where these two types of sensors acquire data. The pink areas are the portions of the EM spectrum that are not blocked by Earth's atmosphere.

Geostationary Weather Satellite Systems

The same geostationary orbital plane, although mostly used by commercial telecommunications satellites, also provides an excellent location for global weather satellite satellites to provide a constant, low resolution, 24-h view of Earth. There is now an operational ring of satellites from several nations that provide a constant, global view of Earth to provide early warning for severe weather. We now take these for granted, seeing them daily in the media. This was a revolutionary capability when the first satellites were launched, and we had the first synoptic view of Earth's surface and weather patterns. The first working weather satellite

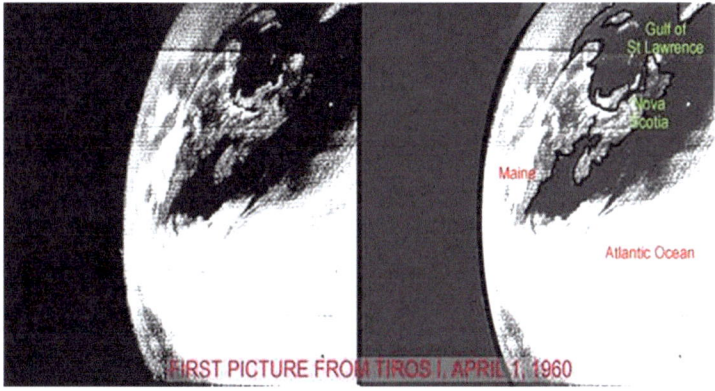

Fig. 5.2 The first TIROS image of Earth in 1960 (Image courtesy of NASA)

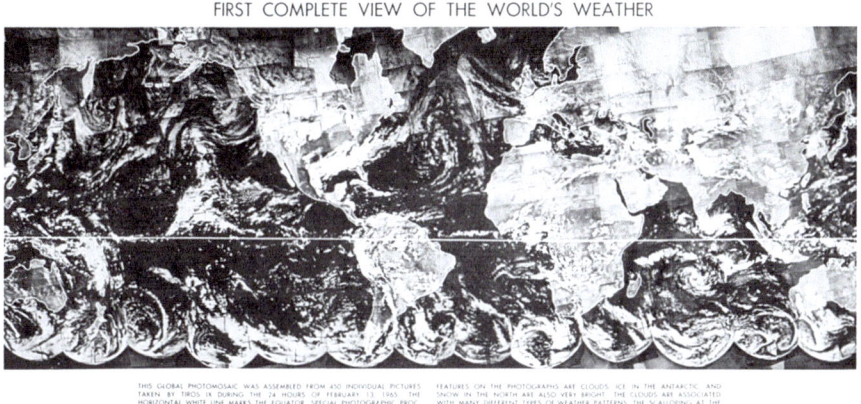

Fig. 5.3 The first complete view of the world's weather in 1965 (Graphic courtesy of NOAA)

was TIROS, launched on April 1, 1960. This pioneering satellite acquired this image of the Canadian maritime provinces, showing the cloud cover (Fig. 5.2).

TIROS also provided the first view of all Earth's weather, in this composite created in 1966 (Fig. 5.3).

The above image was a composite of 450 individual photographs that were processed and combined into a single image by the National Oceanic and Atmospheric Administration.

From these humble beginnings, we have created a global system of weather satellites that are on constant guard for severe weather. There is a ring of satellites, operated by the United States, Europe, India, Russia, Japan and China, which provide a 24-h weather monitoring system (Fig. 5.4).

Fig. 5.4 The GOES full disk image of Earth, by NOAA's GOES-13 satellite on September 4, 2010 (Image courtesy of NOAA)

In the United States, it is NOAA's National Environmental Satellite, Data, and Information Service (NESDIS) that operates the Geostationary Operational Environmental Satellites (GOES). They also provide the associated data to the NOAA's National Weather Service that serves as their primary customer. NOAA-produced satellite data are also broadly available around the world. The two operational GOES satellites provide continual monitoring of the Americas and are used for hurricanes, tornadoes, flash floods, and other weather events. Operating since 1975, the current system has two satellites with overlapping coverage, extending far into the Atlantic and Pacific oceans (or GOES East) and (GOES West) monitors over half of Earth day and night. These satellites provide vital warning for severe weather (Figs. 5.5 and 5.6).

These satellites provide early warning for the development of hurricanes and other severe weather, including a wide variety of events that are vital to the emergency management community ranging from tornadoes to volcanic eruptions (Fig. 5.7).

European and African coverage is provided by the EUMETSAT organization (www.eumetsat.int) from Europe. They operate four METEOSAT first- and second-generation satellites, which are similar to GOES and are positioned over Europe and Africa. India operates the InSat satellites, while Japan operates the Himawari (Sunflower) satellite. China operates the Feng-Yun (Wind and Cloud) satellite, and Russia has operated the GOMS system since 1994. All together, these provide an overlapping constant view of the world's weather (Fig. 5.8).

Geostationary Weather Satellite Systems 71

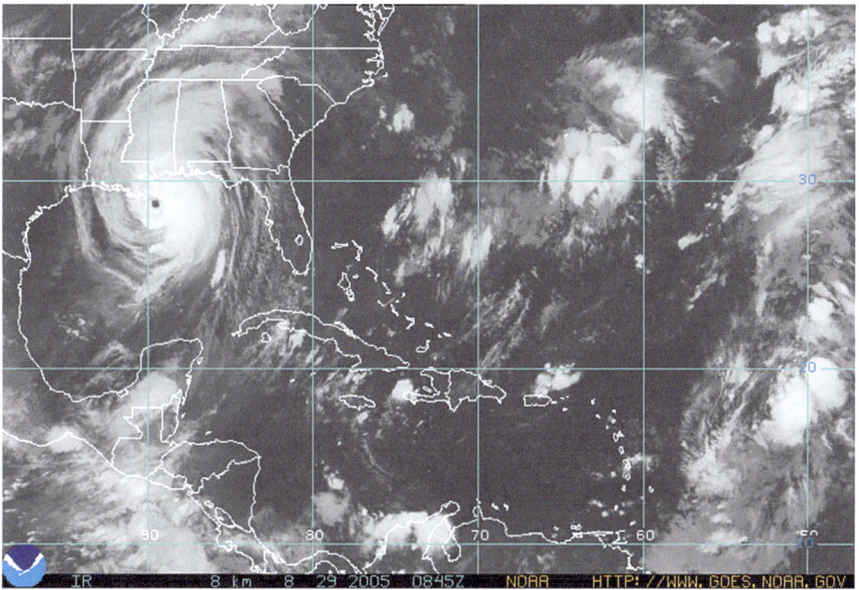

Fig. 5.5 Hurricane Katrina approaching landfall in Louisiana (Image courtesy of NOAA)

Fig. 5.6 Hurricane Bonnie making landfall (Image courtesy of NOAA)

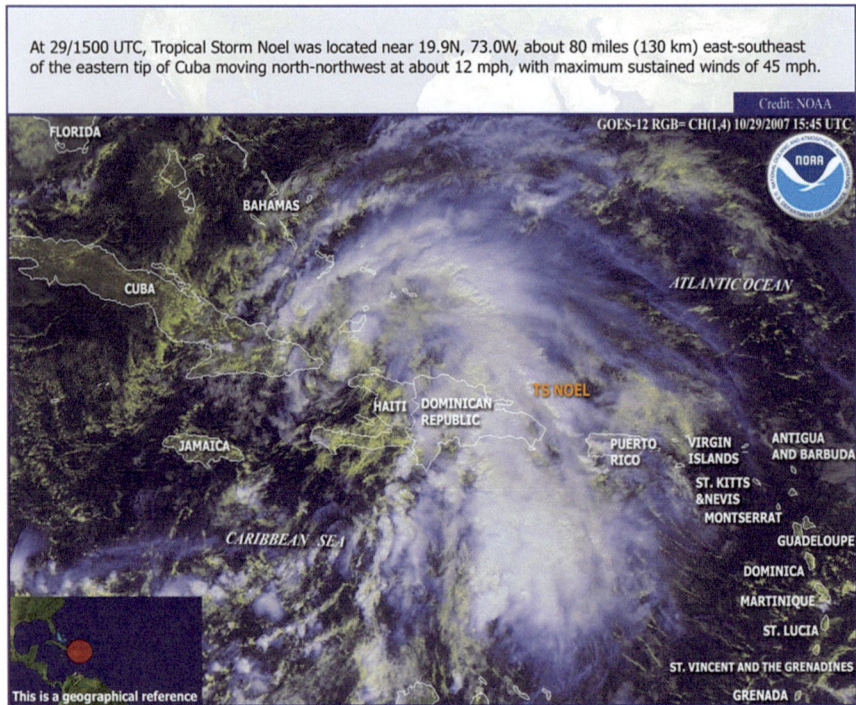

Fig. 5.7 NOAA GOES image of tropical storm Noel (Image Courtesy of NOAA NESDIS)

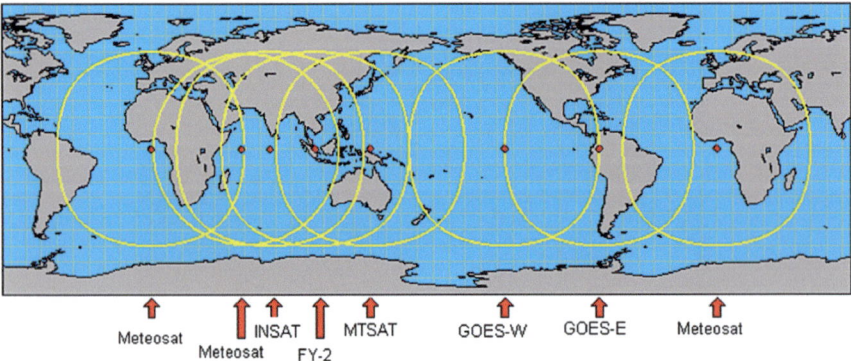

Fig. 5.8 A view of the various GEO weather satellites (Image courtesy of NOAA)

Each of these systems require a significant ground infrastructure of satellite control and operations, and although each system is operated individually there is the coordinating group for meteorological satellites (CGMS) that provides a global observing 'system of systems' and operates the World Weather Watch. The CGMS provides an international forum for the exchange of technical data since 1972 and currently consists of 15 member organizations and six observers. While it does not have staff or legal authority, it provides standards for data so that ship, aircraft, and buoy data can be transmitted.

The World Meteorological Organization of the United Nations (WMO) facilitates worldwide cooperation and standards for weather data and frequency management issues, including a global telecommunication system (GTS) that distributes weather and low-resolution imagery worldwide and that operates the global observation system (GOS). The current list of all current and future GEO and LEO weather satellites, operators, sensors, and other details are maintained at the following website: http://www.wmo.int/pages/prog/sat/satellitestatus.php.

One of the most useful aspects of this work is the distribution of direct-readout satellite-receiving ground stations data from these GEO satellites. This was pioneered over 45 years ago by NOAA, which provided automatic picture transmissions (APT) and high resolution picture transmission (HRPT) capabilities to their GEO and LEO satellites for direct readout on the ground by small receivers. Over 8,000 receivers in over 120 countries have been purchased (or built) by teachers, companies, and government agencies around the world, providing low cost weather data and imagery. These can be connected to personal computers for direct access to weather satellite data for a total of about US$8,000. The new GEONETCAST system (discussed below) significantly reduces this cost.

These satellites also broadcast Weather Fax, high resolution picture terminal (HRPT) and automatics picture terminal (APT) format low resolution weather charts, satellite imagery, and other information that can be received on the ground, in the air, and at sea using a small antenna and laptop computer. Many hobbyists build their own systems from openly available plans. HRPT terminals provide global imagery at 1 km resolution, as well as day and night sea surface temperature, and ice, snow, and cloud cover data. There is no fee or license needed to acquire the NOAA data, while commercial companies provide processed data for a fee. APT systems provide 4 km resolution infrared and visible imagery. Again there is no fee. A list of satellite receiving equipment manufacturers is available from NOAA at: http://www.noaasis.noaa.gov/NOAASIS/ml/manulst.html (Figs. 5.9 and 5.10).

Moderate Resolution Optical Satellite Systems

Both NOAA and EUMETSAT also operate polar orbiting satellites that provide high spatial resolution weather imagery with identical sensors. NOAA operates the polar operational environmental satellites (POES). Each completes 14 orbits per day from an altitude of 830 km (520 miles) and provides a wide range of weather and environmental data, including two daily complete views of Earth (Figs. 5.11 and 5.12).

74   5 Space Remote Sensing Fundamentals and Disaster Applications

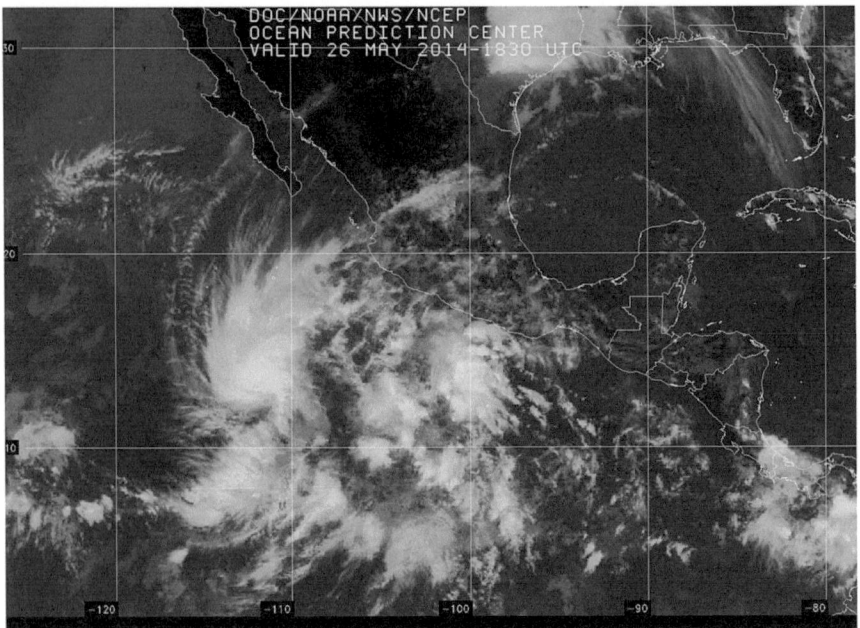

Fig. 5.9 NOAA WEFAX low resolution GOES image (Image courtesy of NOAA)

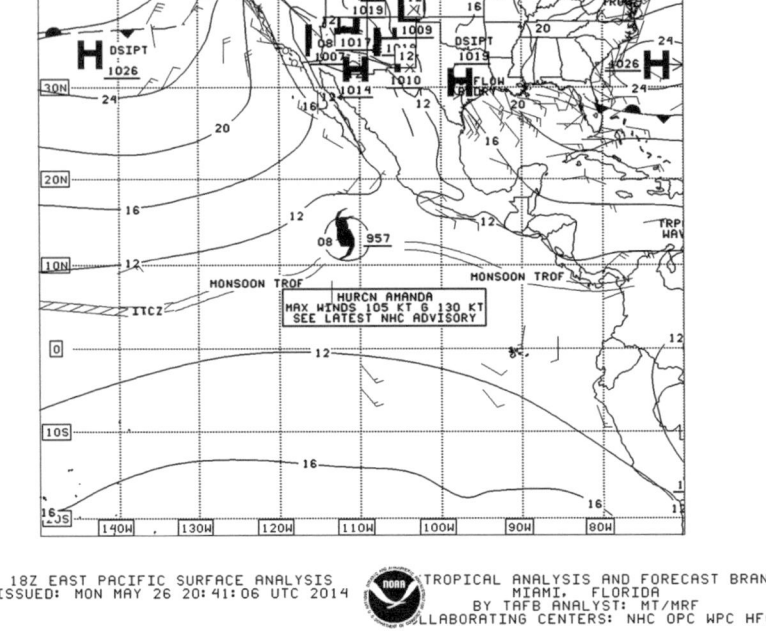

Fig. 5.10 NOAA WEFAX low resolution weather map (Image courtesy of NOAA)

Moderate Resolution Optical Satellite Systems 75

Fig. 5.11 NOAA-19 POES satellite (Image courtesy of NOAA)

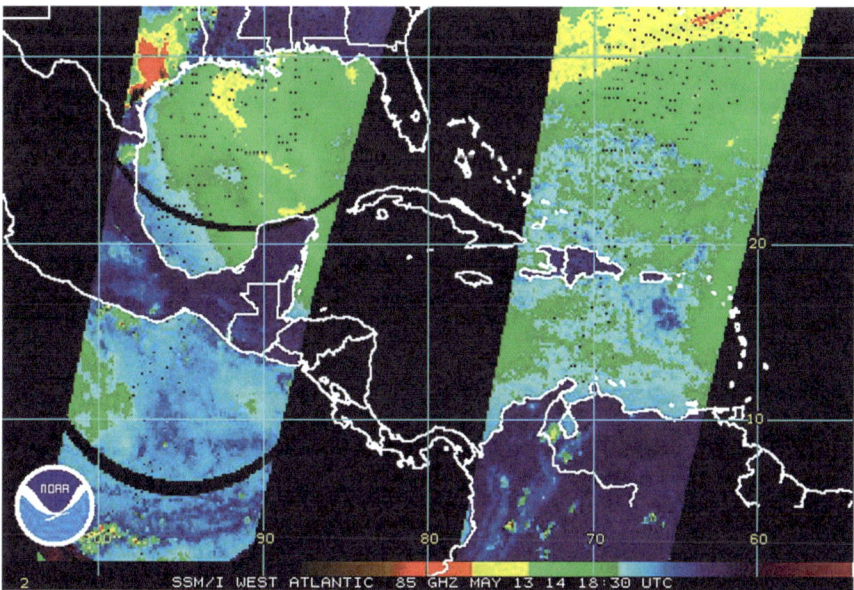

Fig. 5.12 Graphic that shows the imagery from two NOAA POES satellite passes with 85 GHz radiance data (Image courtesy of NOAA)

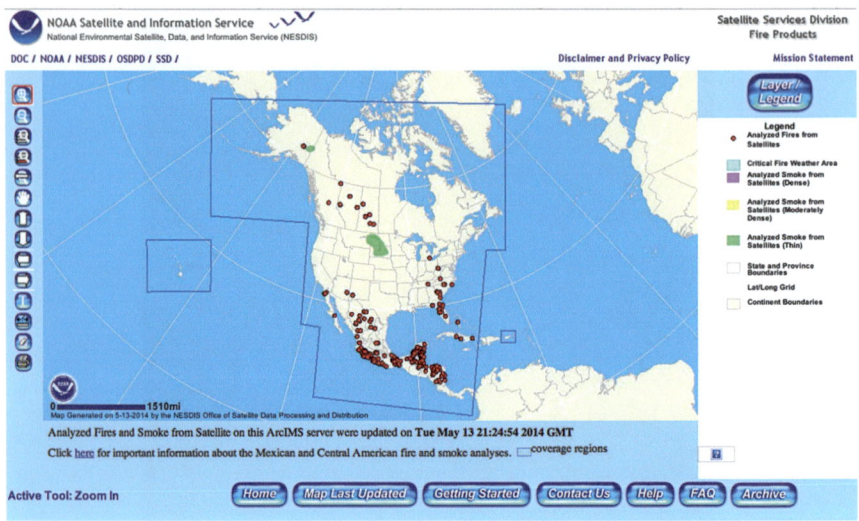

Fig. 5.13 NOAA website with satellite fires and smoke data (Graphic courtesy of NOAA)

NOAA operates a hazard mapping system fire and smoke detection service, and provides the data on its website: http://www.firedetect.noaa.gov/viewer.htm (Fig. 5.13).

The European METOP satellites, operating since 2006, provide data used to create weather models ranging from a few hours to up to 10 days. These two systems (NOAA and EUMETSAT) operate cooperatively, with the U.S. and European satellites flying in certain orbits to maximize mutual coverage.

There are a large number of moderate resolution satellite systems in orbit today. These occupy polar, Sun-synchronous orbits that are at altitudes of between 500 and 1,000 km. These satellites are launched in orbits so as to pass over the same point on Earth every *n* days, depending on the sensor design and altitude of the satellite. The systems provide synoptic, global imagery daily at a much higher spatial resolution than the GEO systems. These moderate resolution range produce imaging ranging from 1 km pixels down to less than 20 m.

NASA operates a large number of polar satellites with direct disaster uses through their EOSDIS, the Earth observation system data and information system. It operates a rapid response capability that provides imagery from 40 different products from MODIS (Moderate Resolution Imaging Spectroradiometer) and other instruments, categorized into ten groups for natural and man-made disasters. (See the MODIS website at https://earthdata.nasa.gov/data/near-real-time-data/rapid-response.) This includes near real-time views of imagery for tracking fires, smoke, floods, dust storms, volcanic ash plumes, and air quality parameters (https://earthdata.nasa.gov/data/near-real-time-data/data/hazards-and-disasters). The fire information for resource management system (FIRMS) provides access to current and archival MODIS fire data. Users can also receive email alerts of fires in an area of interest by subscribing to the free NASA FIRMS fire email alert portal (https://earthdata.nasa.gov/data/near-real-time-data/firms/fire-email-alerts).

The MODIS Sensors

NASA's MODIS is carried on the Terra and Aqua satellites in a 705-km polar orbit. This sensor is able to acquire data in 36 spectral bands every 2 days over the entire Earth with three levels of resolution: 250 m, 500 m, and 1 km. It covers an area some 2,330 km across each pass. It also includes a thermal system and so can determine temperature and therefore fire locations.

Moderate resolution systems such as MODIS are routinely used for disaster work. Its daily coverage, thermal imager (for fires), free availability, and low resolution make it very appropriate for disaster use. One especially useful capability is the thermal infrared (IR) channel, which allows direct observation of fires, with good penetration through smoke (Figs. 5.14 and 5.15).

Fig. 5.14 A MODIS 250-m resolution image of fires and smoke in Baja California (Image courtesy NASA)

Fig. 5.15 A MODIS image of a tornado track that passed through Mayflower, Arkansas, on April 27, 2014 (Image courtesy NASA)

The NASA Tropical Rainfall Measuring Mission (TRMM)

The TRMM is a NASA research satellite designed to measure precipitation in the tropics, and to understand the role of precipitation in shaping tropical weather and climate. It is very useful for a variety of disaster scenarios and can be used to measure extreme weather, including global tropical cyclone, flood, and landslide monitoring (Fig. 5.16).

NASA maintains websites showing global landslide and flooding potential based on TRMM and other data, as shown here (Figs. 5.17 and 5.18).

Landsat

The Landsat satellites, operated by NOAA, have been in continuous operation since 1972, with a spatial resolution of first 80, then 30, and now 15 m per pixel on the ground. Although these satellites offer what is considered moderate resolution, the long history of these data sets allow for excellent time series and before-and-after image comparison. Its wide coverage (approximately 180 km per scene) shows large areas for disasters such as floods and cyclones (Figs. 5.19 and 5.20).

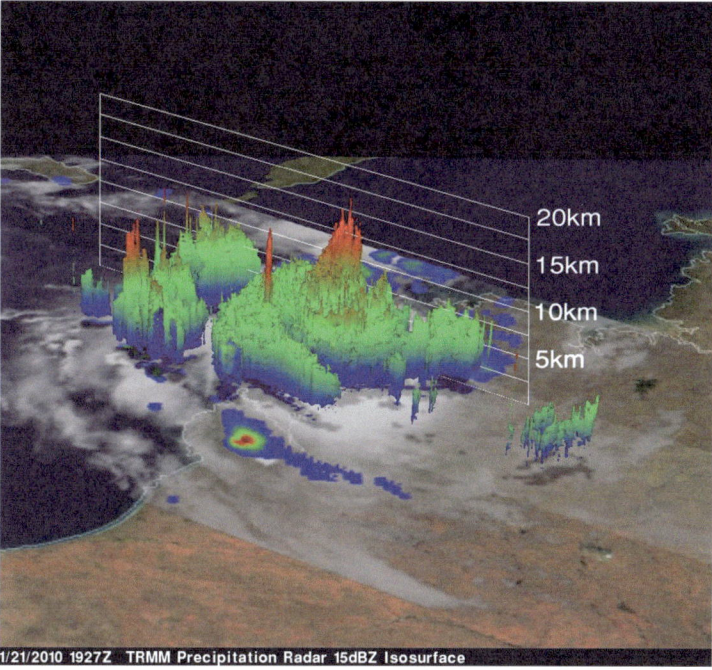

Fig. 5.16 A NASA TRMM precipitation radar image showing thunderstorms near tropical cyclone Magda off the Australian coast (Image courtesy of NASA)

The Disaster Monitoring Constellation (DMC)

There is a constellation of commercial satellites that are specifically designed by Surrey Satellite Technology, Ltd., of the UK, called the disaster monitoring constellation, or DMC. The satellites are designed and built by Surrey, but each satellite is owned and controlled by the individual international consortium members. With five satellites, the constellation covers the entire Earth each day, with spatial resolution ranging from 22 to 2.5 m. The UK, Spain, China, and Nigeria (with two satellites) are the current owners and users, and Turkey and Algeria have previously had satellites in the constellation. The constellation became operational in 2003 and has since provided daily imaging of all of Earth with moderate resolution. It was updated in 2009 with the current 22-m capability. All data are provided within the disaster charter without cost when there is an incident or a disaster.

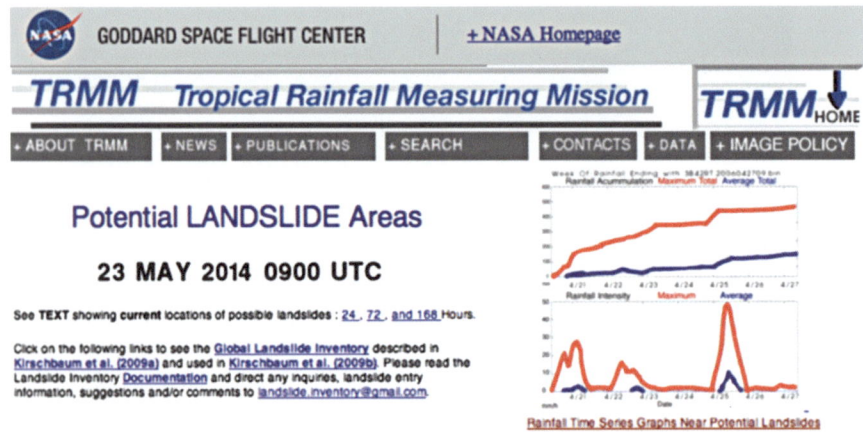

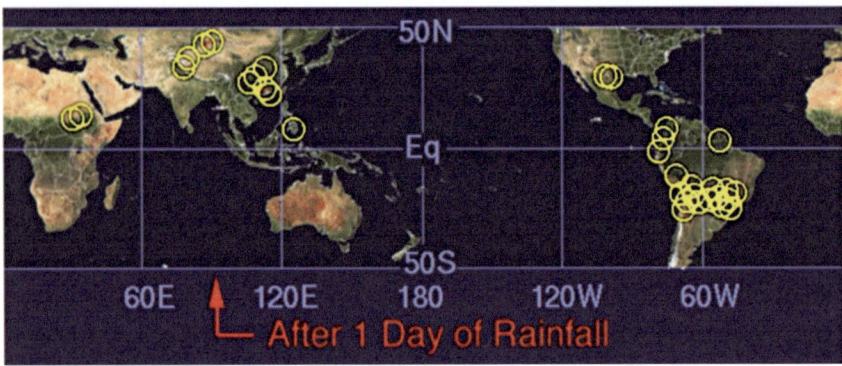

Fig. 5.17 TRMM global data indicating potential landslide areas (Image courtesy of NASA)

High Resolution Optical Satellite Systems

There are several new satellites that provide ultra-high spatial resolution – that is, below 1 m pixels on the ground. The current maximum commercially available resolution in the United States is 30 cm pixels, comparable to aerial photography in detail, but with a small swath only a few kilometers wide. These systems provide detailed coverage over any cloud-free area of the world, allowing individual features such as roads, bridges, and houses to be clearly identified. They provide excellent capability for disaster management and response, including preliminary and detailed disaster assessment. This allows the status of infrastructure, bridges, and individual structures to be determined quickly and accurately from space. Before and after imagery provides the basis for these assessments by comparing imagery either manually or, experimentally, by automated processing of the data.

DigitalGlobe (www.digitalglobe.com) operates a fleet of satellites that provides spatial resolution imagery at a resolution of between 1.7 and 3.2 m in multispectral

Developing Technologies

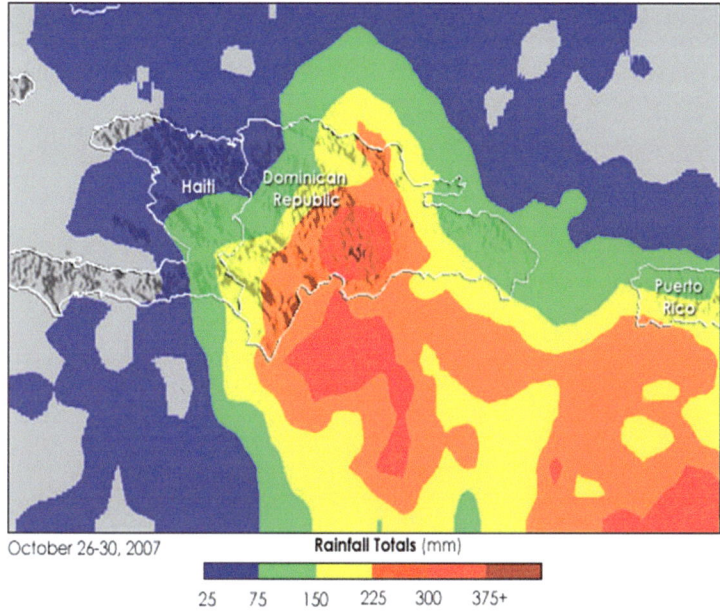

Fig. 5.18 A TRMM image of rainfall totals over the Dominican Republic, showing total rainfall accumulations (Image courtesy of NASA)

bands and resolutions varying between 31 and 82 cm for panchromatic, color, and color infrared spatial resolution. Digital Globe Satellites have a swath width of between 11.3 and 18 km per pass. The imagery is very high quality and has been used in a wide variety of disasters around the world.

Developing Technologies

The space remote sensing world continues to evolve quickly, and there are several new systems available now or coming soon. Skybox Imaging (http://skybox.com) is launching a constellation of 24 satellites that will provide 1-m detailed imagery. Its first SKYSAT-1 is now operational, and it is using a subscription and data stream model, including data analytics that can be very helpful in disaster assessment. Skybox also is providing the only HD video capability from space, with up to 90 s of video per pass. This could provide vital information regarding damage and refugee movement or other services that are highly time dependent or involves motion.

Planet Labs is another recent start-up (http://www.planet-labs.com) that will operate the world's largest fleet of small remote sensing satellites. They currently have the first 28 small and low cost satellites in orbit, each run by Android phones onboard that use low cost hardware that is not space qualified. They have funding

Fig. 5.19 Mississippi river floods in April and May 2011 shown before and after on *Landsat 5* imagery (Image courtesy of NASA)

for another 72 satellites, which will make a constellation of 100 satellites, imaging the entire Earth each day with 3- to 5-m resolution imagery.

An interesting new data source has recently become available from JAXA, the Japanese space agency. JAXA is creating imagery from two HDTV video cameras installed on the Kibo module of the International Space Station. Video movie data, as well as geo.TIFFs, are available and are useful for wide area monitoring.

Passive vs. Active Remote Sensing Systems

Although these passive remote sensing systems are very useful, they all share the common limitation of not being able to be used through clouds, smoke, or at night. But there is another class of remote sensing satellites that can easily acquire imagery at night and through rain and clouds, which make them particularly useful in disaster applications. These are active remote sensing systems, such as RADAR. They are fundamentally different from passive systems such as MODIS or Landsat, in that, rather than simply measuring light energy reflected off Earth's surface, they send out their own burst of electromagnetic energy and measure the energy reflected off the target. This allows these systems to acquire imagery at night and through clouds or smoke, even in driving rain (Fig. 5.21).

Passive vs. Active Remote Sensing Systems

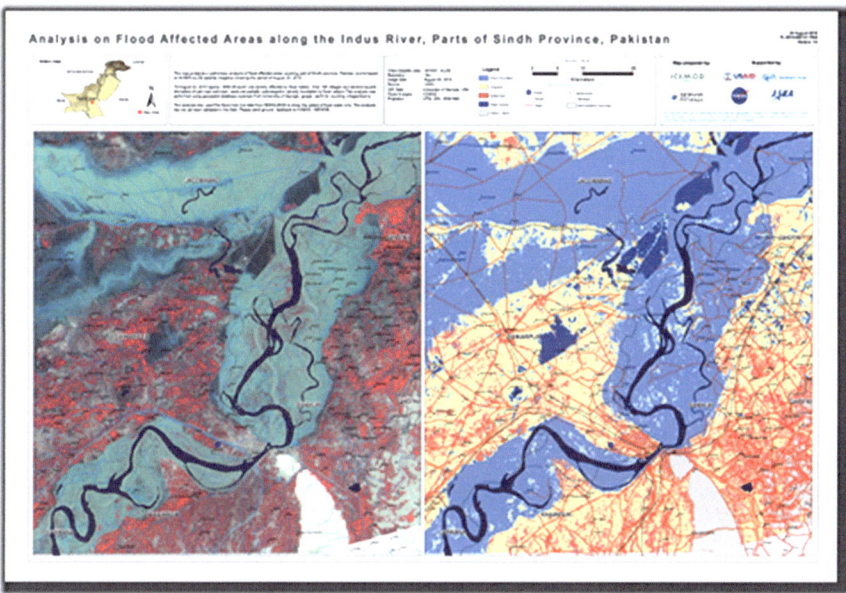

Fig. 5.20 NASA Landsat image (*left*) and derived map (*right*) of flooding in Pakistan based on a Landsat image (Image courtesy of NASA)

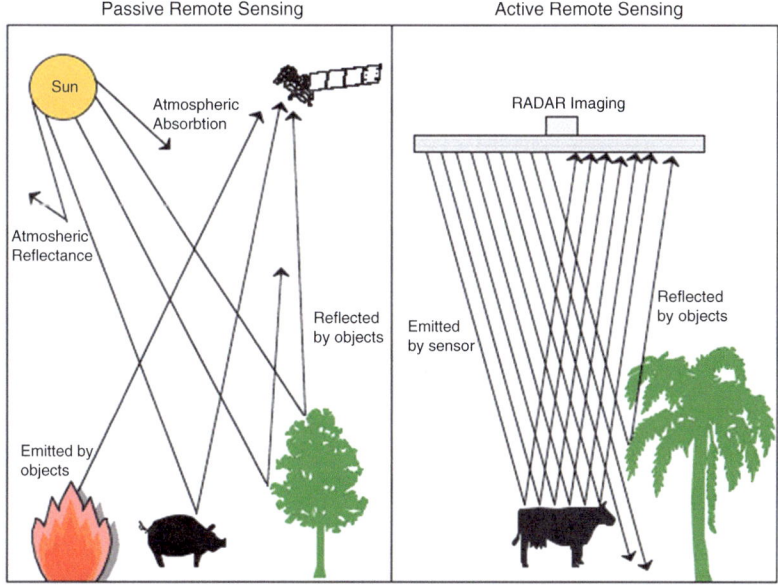

Fig. 5.21 Passive vs. active remote sensing systems (Image courtesy of NOAA)

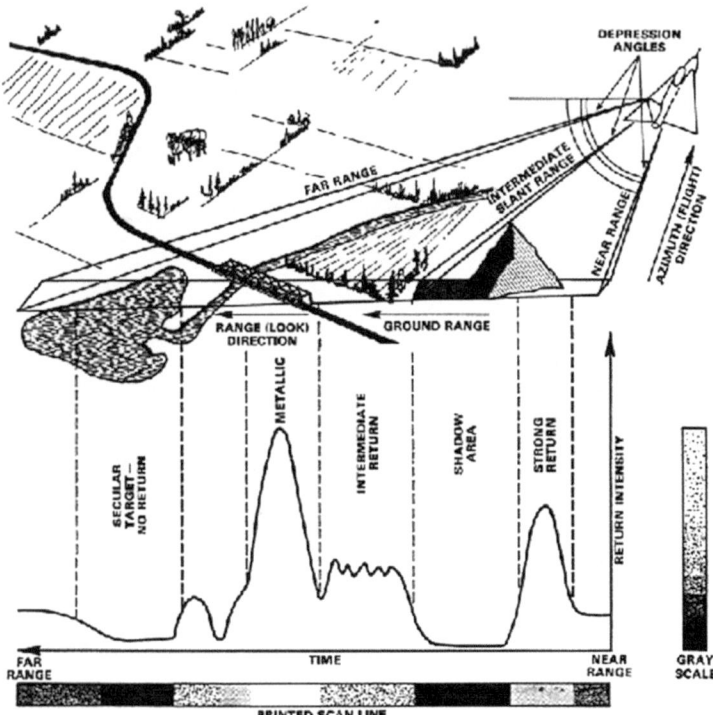

Fig. 5.22 Diagram showing the concept of RADAR-based active remote sensing (Image courtesy of NASA)

Fundamentals of Active/RADAR Remote Sensing

A RADAR remote sensing system sends its own burst of electromagnetic energy in the microwave region of the spectrum from the satellite's antenna down to Earth's surface at an angle. Then, using the same antenna, it records the amount of energy that is reflected back, the date of which is stored and transmitted down to the ground for processing and analysis. Different types of land and water will absorb or reflect back different amounts of energy, and this allows us to accurately map the landscape. RADAR satellites and data are more complex than passive systems, and the results do not look like passive images until they are processed by dedicated computer systems, but they provide a powerful and complimentary capability to passive systems such as MODIS and Landsat (Fig. 5.22).

When a burst of energy is sent down from a RADAR satellite at an angle from the platform, the reflected image that returns is recorded as stronger or weaker reflections, depending on the composition of the landscape.

RADAR imagery varies widely in both the wavelengths and spatial resolution. All RADAR systems operate in the microwave region of the spectrum, with much longer wavelengths than optical systems. Data ranging from 30 m down to less than 1 m are now available worldwide. The European TerraSAR-X system, for example, has five modes, with flexible spatial resolution ranging from .25 m, 1 m, 3 m, 18.5 m, to 40 m, with scenes ranging from 4 km to over 270 km wide.

The Canadian Radarsat 1 satellite was very capable and was followed by the Radarsat 2 system, which was launched in 2007 and is still in operation today. The next generation system, Radarsat Constellation, is in development and will consist of a three-satellite constellation to be launched in 2018. It will provide complete coverage of Canada's land and oceans on a daily basis, with disaster management being one of the three main designed uses. Applications will include predicting flash floods through analysis of snow accumulation and melt, wind speed analysis for severe storms, soil moisture for forest fire analysis, oil spill analysis, pipeline monitoring, and more. It will be able to measure subtle, 1 cm changes in surface position, useful for earthquake analysis and more. The COSMO-SkyMed system is a four-satellite constellation that consists of identical synthetic aperture radar (SAR) sensors funded by the Italian space agency and the Italian military. These systems have very high resolution military modes, but also have several modes ranging from 3 to 100 m that are useful in a variety of disasters and can collect up to 450 individual images per satellite per day. All of these RADAR systems, and more, are available through the disaster charter (discussed below) and are routinely used for disaster response (Fig. 5.23).

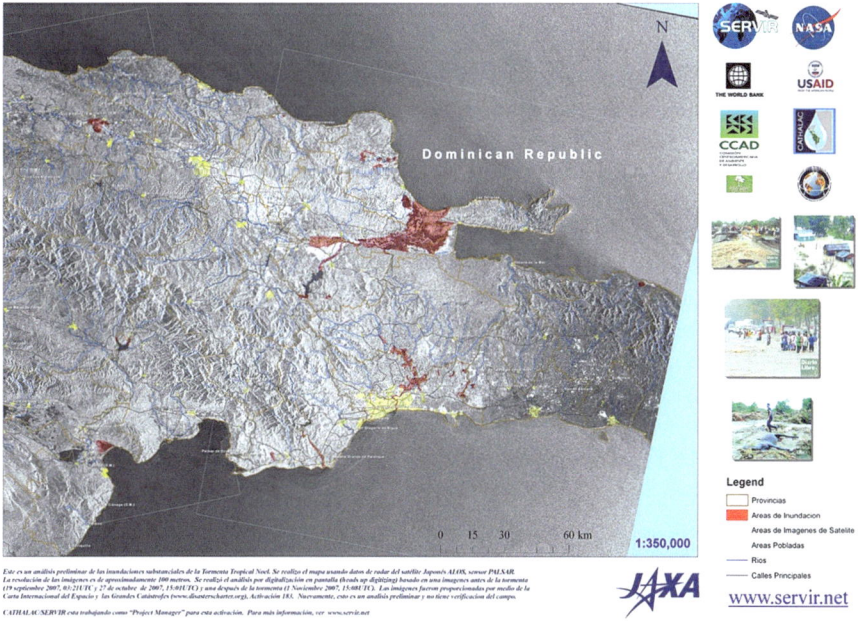

Fig. 5.23 This image from a JAXA RADAR satellite highlights flooding in the Dominican Republic (Image courtesy of JAXA)

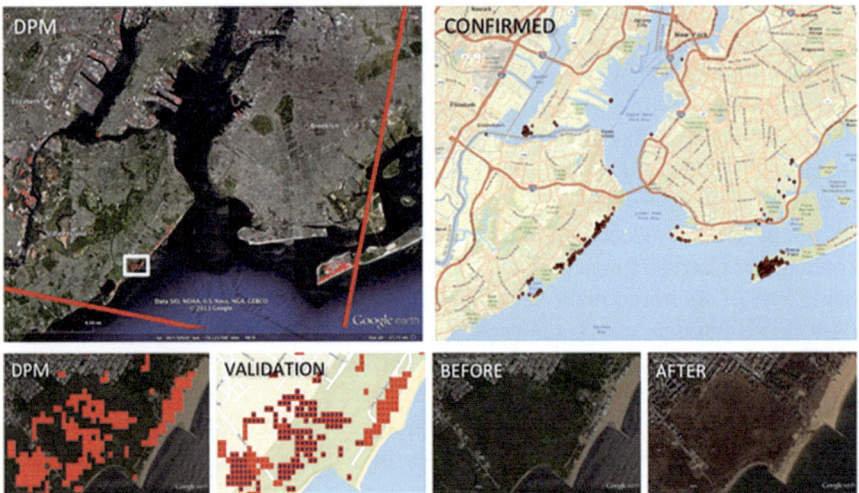

Fig. 5.24 ARIA image of damages caused by Hurricane Sandy in New York City area using X-band RADAR from the COSMO-SkyMed satellite constellation

RADAR imagery requires significant data processing and time to create images that can be interpreted for disaster uses. The NASA JPL/Caltech advanced rapid imaging and analysis (ARIA) center (http://aria.jpl.nasa.gov) seeks to provide near real-time image processing to improve disaster response and situational awareness. Products include SAR interferograms, surface displacement maps, horizontal and vertical ground displacement, damage proxy maps, and critical infrastructure assessment. These maps are produced in only 1 day (Figs. 5.24 and 5.25).

This JPL image shows an area 40 by 50 km. It was processed by NASA from data received by the Italian space agency's COSMO-SkyMed satellite. The detailed images were achieved using NASA's ARIA center. Red areas show heavily damaged areas.

Synthetic aperture radar systems (SAR) can acquire all weather, day or night, rain or shine imagery, which is particularly useful for disaster applications. These allow acquisition of data through clouds and provide a wide range of information about flooding, vegetation, and soil moisture. By combining two RADAR images, it is possible to create digital elevation models (DEMs) for use in a GIS system, with very high precision. This process, referred to as interferometry, allows vital data about earthquakes, volcanoes, landslides, ground sinking, and ice movement.

The colors in Fig. 5.26 show ground displacement differences between two RADAR images before and after the earthquake.

Fundamentals of Active/RADAR Remote Sensing

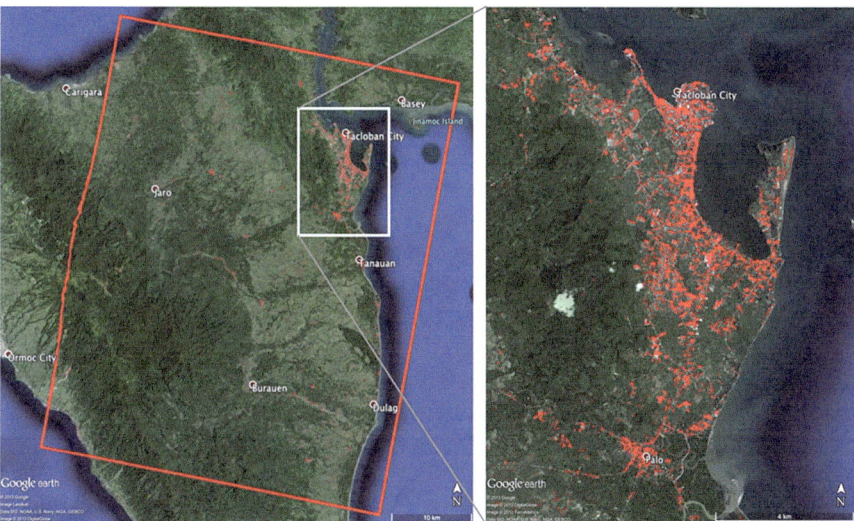

Fig. 5.25 NASA processed image of the effects of super typhoon Haiyan in the Philippines on November 8, 2013 (Image courtesy of NASA)

Fig. 5.26 A satellite-based interferometry map of an earthquake (Image courtesy of NASA/JPL)

Aerial Imaging Systems and Unmanned Drones

There is a new class of systems that can be very useful in disaster response, consisting of unmanned aerial drones and unmanned aircraft of many types. Keep in mind that disaster managers really do not care where the data come from as long as it is useful to them, and aerial drones can provide excellent damage assessment data if the weather allows. Benefits of these, compared with satellite imagery, include the ability to control exactly when the data are acquired, higher spatial resolution, immediate acquisition of the data, and more. The first use of these was in Bosnia in 1994, and drones have been used around the world in various ways since then.

There is a wide range of drone types that are available, ranging from sophisticated and expensive military systems costing millions of dollars to small, hand-launched quad-copters that can fly only meters above the ground and cost only a few hundred dollars.

At the very high end (literally) is the U.S. Air Force RQ-4 Global Hawk, which has been flown in support of multiple disasters since 2007, including California forest fires, Haiti after the 2010 earthquake, the Japan tsunami in 2011, and Typhoon Haiyan in the Philippines in 2013. In the Philippine incident, the drone was pre-positioned and flew its first mission – acquiring detailed imagery of the affected area – only 2 h after it was requested. It flew three missions and acquired over 50 h of data, including over 300 images that were made available to the disaster response team. Its first use in the United States in 2007 raised several issues concerning privacy and civilian airspace regulation with the FAA, but it demonstrated the potential of these systems, and these craft will continue to be used in disaster response due to their long range (over 8,000 km), 30-h endurance and high-quality remote sensing and satellite telecom systems (Fig. 5.27).

The smaller MQ-1B Predator has also been used in responding to forest fires in California and proved useful in providing real-time thermal video of the situation to the firefighters.

Smaller, commercially available drones are now available and range in cost from US$3,000 to $50,000. These smaller drones can be carried in a car or vehicle, be launched by hand, and acquire real-time data or high resolution imagery for later analysis.

The largest limitation, in the United States, to the use of drones in disaster response is the legal status of these remotely piloted vehicles to fly and especially the policies of the Federal Aviation Administration. Currently, any civilian or military drone flying 400 feet (122 m) above the ground must get a special waiver from the FAA, a process that takes, at best, 60 days. Even though there is an exception for disasters, the many limitations make it so hard to obtain that they are rarely sought. There is also often a no-fly zone placed over disaster areas to limit aerial tourists and 'sightseers.' The U.S. Congress has required the FAA to come up with new, more workable policies by 2015, so this situation may change.

Fig. 5.27 U.S. Air Force RQ-4 Global Hawk drone taking off (Image courtesy of the U.S. Air Force)

The American Red Cross, for example, had access to a drone and had conducted training exercises for using it for disaster assessment, but could not use it in the 2013 Moore tornado because of a no-fly zone and the FAA's refusal to allow its use. The same restrictions also applied in the case of Hurricane Katrina.

These restrictions are less stringent in the developing world, and we will likely see more use of these small aerial systems in the future.

Developing Remote Sensing Technologies

At the very smallest size, micro unmanned aerial vehicles (or MUAVs) have been proposed as an excellent way to acquire contamination information in disasters involving chemical, biological, or nuclear materials dispersed into the atmosphere. The rapidly evolving technologies of micro-aerial vehicles, micro-sensors and related systems are at the extreme low end of the disaster remote sensing response, but these will continue to be developed. It is likely that new uses will evolve. One application might be to enter dangerous structures for search and rescue or monitoring dangerous situations such as nuclear discharges without directly endangering responders. Papers have been presented that have proposed the use of micro drones to deliver medical supplies or food in disaster areas, and the technologies to do this will continue to develop.

Chapter 6
Precision Navigation and Timing Systems

A relatively recent development in satellite applications is the enormous growth in the use of positioning, navigation and timing (PNT) satellite systems such as the U.S. Global Positioning System, or GPS. There are now several similar systems, so they are collectively referred to as PNT systems, for precision navigation and timing systems or global navigation satellite systems (GNSS), which is the term used by the International Telecommunication Union. These PNT systems use a constellation of satellites to provide very precise timing, navigation, and positioning data globally, and they have valuable applications for disaster management.

PNT Systems

The U.S. GPS system was the first and is still the most commonly used satellite navigation system. It is made up of three segments: the space segment (satellites), the control segment (ground stations and control centers), and the user segment (your smart Phone, such as an iPhone or Android phone, or hand-held GPS receiver) (Fig. 6.1).

The GPS space segment (also known as Navstar) consists of six orbital planes inclined 55° at an altitude of 20,200 km, so that each satellite orbits Earth two times per day. There are a total of four to six satellites in each plane, so that a minimum of four satellites are always in view anywhere on Earth. The constellation, or 'bird cage' as it is sometimes called, required a minimum of 24 satellites, but often it has up to 32. Frequently many more are visible, and the satellites are constantly rising and setting above you. The system is managed for the U.S. government by the U.S. Air Force and has been operational since 1978, with global coverage since 1994.

The GPS system is actually a very precise timing system, and each satellite has several atomic clocks and constantly broadcasts a coded message stream that repeats itself. Your receiver processes the different signals to determine your

Fig. 6.1 The GPS constellation of satellites (Image courtesy of GPS.gov)

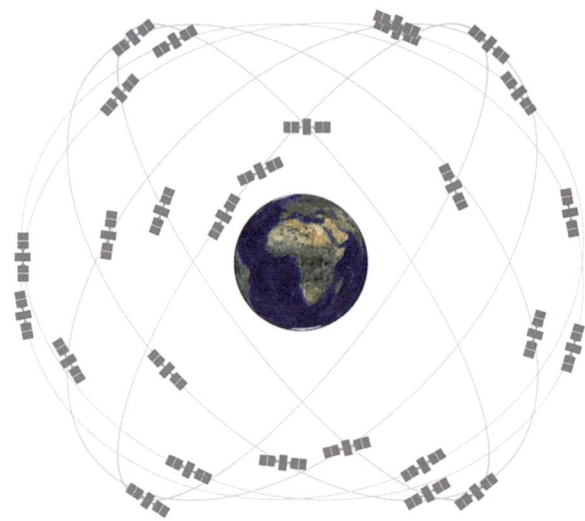

local position and time. The GPS control segment includes the master control station at Schriever Air Force Base in Colorado Springs, Colorado, and alternately, 12 command and control sites and 16 monitoring stations located around the world. This gives the USAF 2nd Space Operations Squadron daily operational control over the space segment. The monitoring stations track the satellites as they pass overhead and pass the data back to the master control station. Four control stations, located in Kwajalein Atoll, Ascension Island, Diego Garcia Island, and Cape Canaveral, are used to communicate with the satellites and to update navigation and other data to them.

The user segment is your GPS receiver. There are a wide range of GPS receivers available, ranging from inexpensive hand-held units to car navigation systems with moving map displays to high end surveying systems with sub-centimeter accuracy (Fig. 6.2).

The U.S. government is committed to continuing to provide GPS to the civilian community for the foreseeable future. The current objective for civilian service is to meet the positioning service performance standard. This is to provide a worst case accuracy of 7.8 m 95 % of the time. In practice, the actual accuracy depends on many variables, including receiver quality, atmospheric and ionospheric conditions and blockage due to buildings and terrain, but is often much better. This accuracy is augmented in the United States by the Nationwide Differential GPS system (NDGPS), a ground-based augmentation system that also provides system integrity data. These broadcast a local correction signal around a limited area. Some 50 countries around the world operate similar systems to a common international standard for interoperability. Over broader areas, the satellite-based Wide Area Augmentation System (WAAS) provides continental-scale augmentation. WAAS provides highly accurate navigation primarily for aircraft, but the system also works for hand-held receivers (Fig. 6.3).

PNT Systems 93

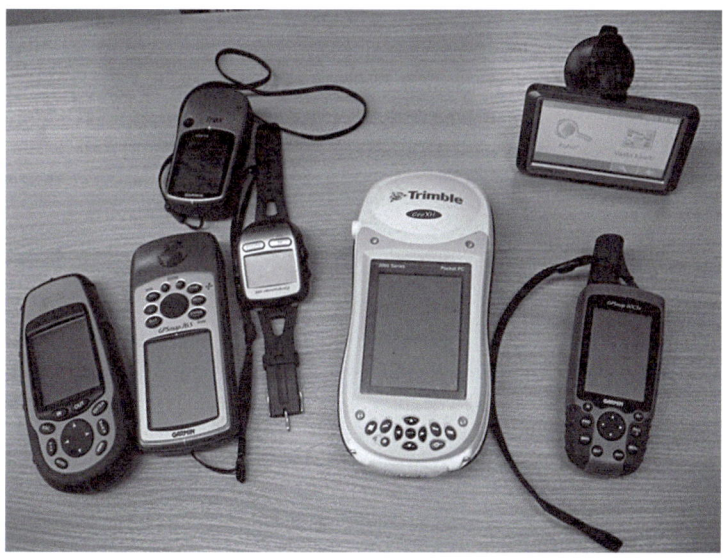

Fig. 6.2 Common hand-held and auto GPS receivers (Image courtesy of Wikimedia Commons)

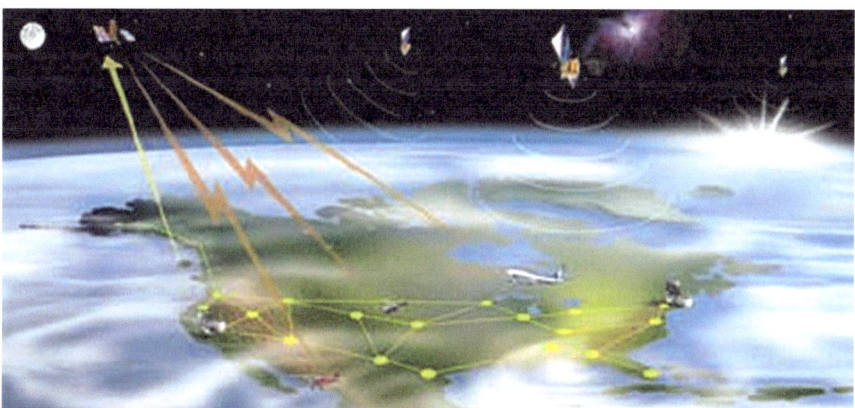

Fig. 6.3 The WAAS system uses commercial telecommunication satellites to broadcast augmentation services for the GPS constellation (Image courtesy of the FAA)

The WAAS system operates in North America, and there are several compatible international systems, including the European Geostationary Navigation Overlay Service (EGNOS), the Japanese Multi-functional Transport Satellite that provides the Satellite Augmentation System Service (MTSAT-MSAS), and the Indian GPS and Geo-Augmented Navigation (GAGAN) system. Many inexpensive hand-held GPS receivers are WAAS-enabled for increased accuracy.

Russia operates the Glonass GNSS system, which is very similar in design to the U.S. GPS system. It is the only fully operational GNSS system today in addition to the U.S. GPS. Many receivers today, including millions of Apple iPhones, have dual GPS and Glonass capability that improves overall accuracy and reliability. Galileo is the European system, which is not yet operational but well on the way. It will be, again, very similar to the GPS and Glonass systems and will have both an open and encrypted channel with different levels of precision and will essentially be similar to GPS and Glonass in design and operation. Japan is deploying what is known as the Quasi-Zenith or Figure 8 Satellite System that provides navigational and communications services. China is developing the Beidou/Compass system, which consists of a combination of GEO and MEO satellites and will be operational over the Chinese and Asian territories. Finally, India is developing the Indian Regional Navigation Satellite System (IRNSS), which will be similar to the Chinese Compass system for regional use around India. The additional systems by Japan, Europe, China, and India will be deployed over the next 5 to 6 years and will provide important redundancy to the GPS and Glonass systems.

Disaster Applications

GPS and other PNT satellites are becoming a critical component of disaster work, providing both timing and location data for search and rescue teams, damage assessment, mapping, fire mapping of hot spots, storm tracking, and flood prediction, among others. A key practical use of GPS in disasters is emergency vehicle fleet management, often referred to as vehicle tracking systems (VTS). These are where GPS as well as other systems can provide real-time tracking of all ground vehicles, boats, and aircraft, and even relief supplies involved in the relief effort. They allow the disaster management teams to view their resources and watch the disaster response unfold. They can more efficiently track and dispatch their resources and more effectively manage the overall response. Such systems can also provide additional information such as engine RPM, emergency lights on or off, fuel and water amount, etc. There are a wide variety of commercial GPS tracking systems available, and several new Open Source systems as well. These generally consist of a GPS device, a transmission device (radio or cell phone), a tracking server that receives the data, and a user interface, which is often a web-based browser (Fig. 6.4).

GPS can also be very useful for field data collection. The IkeGPS, as an example, is an innovative, integrated GPS field data collection system. It consists of a single, hand-held, and ruggedized device with a GPS capability of sub-meter accuracy. It also includes a 5 megapixel digital camera, digital compass, laser range-finder, and mobile computer. This system can be used for very rapid and accurate data collection and damage assessment in the field. The laser range finder allows the user to remotely collect position data of features that cannot be safely accessed, up to 1,000 m away (Ikegps.com) (Fig. 6.5).

GPS In-Situ Networks

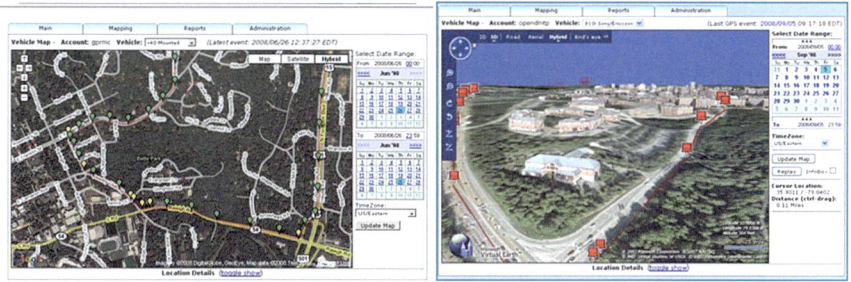

Fig. 6.4 An example of an open source real time vehicle tracking system (Image supplied by the author)

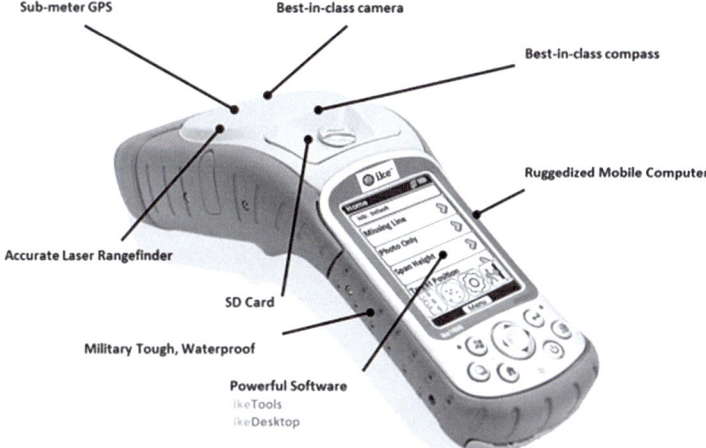

Fig. 6.5 IkeGPS hand-held integrated data collection system (Image courtesy of Ikegps)

GPS In-Situ Networks

There are a growing number of in-situ weather, river height, and other remote monitoring systems, and there will be much more of these in the future. Large networks of distributed data collection platforms can serve multiple functions, including early warning and real-time broad data collection before, during, and after an event.

In 2013, JPL and the Scripps Institution of Oceanography at UC San Diego announced the new GPS-based real-time earthquake analysis for disaster mitigation network to provide enhanced warning for earthquakes, tsunamis, and other extreme events in the western United States. The system uses real-time data from existing GPS monitoring stations that have been upgraded with low-cost seismic and weather data sensors (Fig. 6.6).

Fig. 6.6 GPS station (Image courtesy of NASA/JPL)

This system will warn of earthquakes, track flash floods, and assist in damage assessment for critical infrastructure in near real time. It is based on integrating GPS, accelerometers, pressure and weather data – all collected in real time throughout a region. Data generated can include ground motion, for example, which could provide early estimation of damage. Buildings could automatically disable elevators in earthquakes, for example, as soon as a warning is received. Bridges could also detect deformations and automatically stop traffic.

Hundreds of high-quality GPS units are already located throughout California, which can provide both earthquake intensity as well as water vapor data, which is now constantly collected through the atmospheric distortion that water vapor causes in the GPS data, rather than through twice daily balloon launches.

Figure 6.7 shows the make up the real-time earthquake analysis for disaster mitigation network in California. Red areas represent the locations that have the largest probability of earthquakes. These new, upgraded systems can detect the very fast-moving 'P' waves of an earthquake, making possible automatic warnings for the later arrive of the slower, but more intense 'S' waves, which do the most damage. Such systems could provide warning between several seconds to a few minutes, depending on the distance to the epicenter of the earthquake, and could

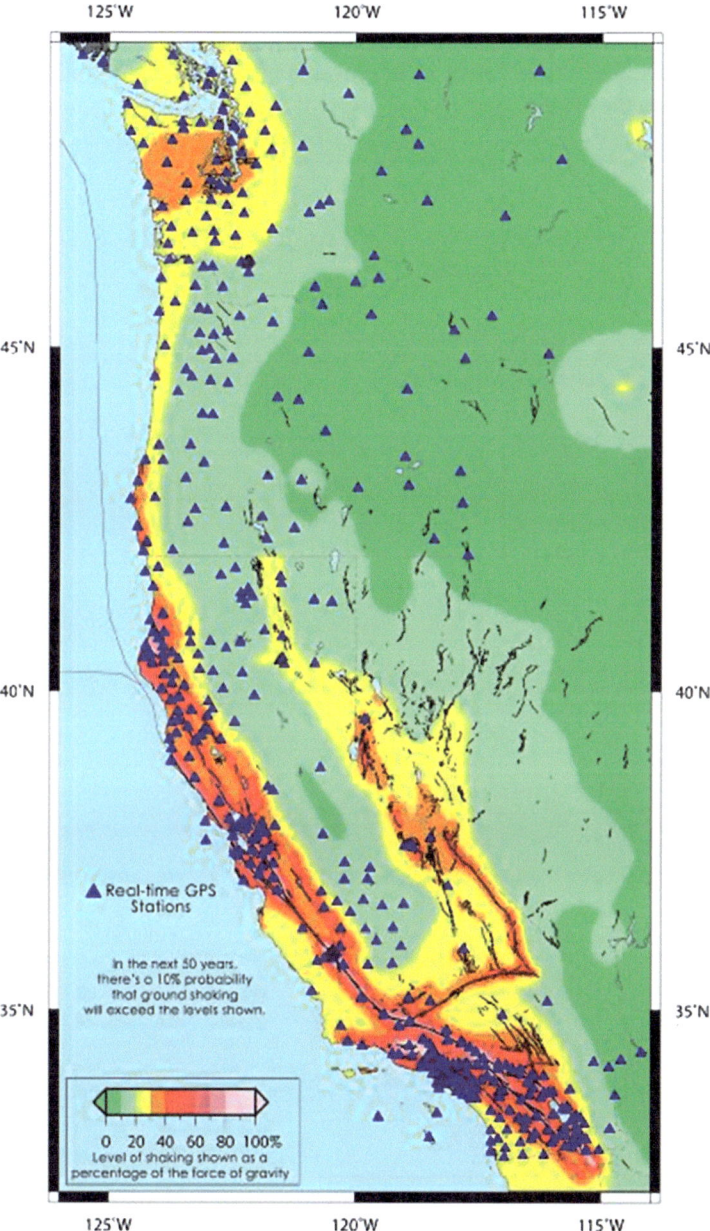

Fig. 6.7 Map of the location of over 500 real-time GPS monitoring stations (Image courtesy of USGS)

automatically trigger warnings, notify emergency personnel, disable elevators, close bridge access, etc. throughout a large area. Existing rain gauges can now easily be equipped with Internet or microwave broadcasting, to provide real-time flood alerts in the same way.

Chapter 7
Geographic Information Systems

Although not technically space systems, computer software programs called Geographic Information Systems, or GIS, are at the heart of disaster management data integration, analysis, and sharing. GIS is where all the information from satellites, GPS, weather RADAR, field damage assessments, paper maps, and other sources are integrated and displayed for analysis and action. Many international, national, state, and local disaster organizations and others are now using GIS systems and are also routinely distributing their data in GIS formats so that it can be directly incorporated into GIS systems.

Figure 7.1 indicates the various layers of a GIS display. This shows the real landscape at the bottom, with various raster imagery and vector map data above. The GIS is sufficiently flexible that any type of spatial information can be integrated into it.

Critical GIS Data Layers

There is a wide range of what are referred to as critical data layers that are vital in all phases of a disaster cycle, and which need to be created well before a disaster strikes. These can include, but are not limited to:

- Boundaries – local, regional, national, international
- Chemical manufacturing facilities, toxic release sites, hazardous materials storage/nuclear storage facilities
- Education – schools and libraries
- Emergency services – fire, police, EMS, public works, seats of government, continuity of operations sites, prisons, other secure facilities
- Energy – gas storage and processing, power plants, pumping stations
- National symbols – public venues
- Tourism and cultural heritage (archaeology and history)

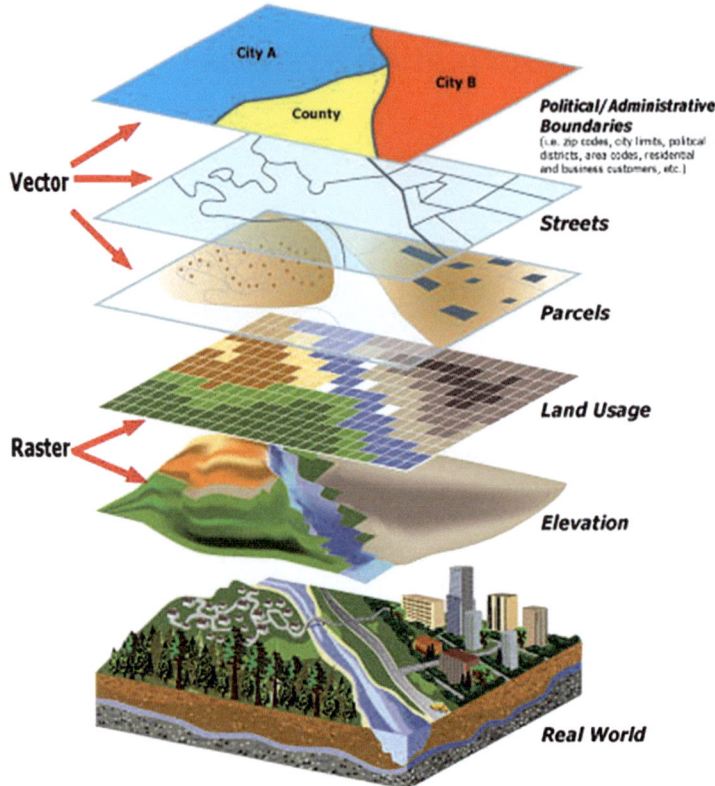

Fig. 7.1 A geographic information system display (Image courtesy of NOAA)

- Public health – hospitals, pharmacies, blood banks, vet clinics, morgues, and mortuaries
- Potential shelter sites, mass feeding sites, decontamination sites
- Special areas (military, prisons)
- Weather – sirens, weather service sites, monitoring sites
- Dams and other potential hazards – flood zones
- Housing – major housing zones, hotels, care facilities
- Transportation – roads, airports, bridges, buses, helipads and landing sites, harbors, evacuation routes
- Logistics – Existing and potential logistics staging sites, warehouses, responder housing, office space, etc.
- Utilities – power, cell, wireless, TV, electrical stations and system, oil pipelines, Gas, telecom
- Other – national security, etc.

Many of these are the standard datasets that most governments acquire on a routine basis and that can be used for many other applications. In addition to these, disaster managers require additional data on what is termed sensitive infrastructure that is specific to this application. These layers can include the location of emergency facilities and shelters, emergency operation centers, continuity of operations sites, logistical warehouses, public safety locations, nuclear, chemical, and biological materials storage sites and other potential hazards, telecommunications facilities, mortuary facilities, evacuation routes, and more. These data should be, as much as possible, developed in advance, and should be frequently updated and appropriate metadata should be completed so that all users know the source and date of each data layer. It is vital that back-up copies are made frequently, and that independent back-up copies are kept off site. Constant updating will be required in an actual disaster, as the pieces are moved around the chess board on a daily basis.

The term common operational datasets (CODs) refers to data that are used by all organizations across the response. The term fundamental operational datasets (FODs) refers to data that are used by a specific part of the response, such as security or feeding. The U.N. Inter-Agency Standing Committee (IASC) has established guidelines on common operational datasets in disaster preparedness and response to help national authorities and humanitarian organizations establish common approaches to the creation of appropriate data, data access, which organization should be the caretaker, how often the data should be updated, etc.

In major U.N.-managed responses, where many individual organizations are participating, a 'Who does what where? (3Ws)' map is frequently produced, showing which organizations are providing services in which areas.

In the United States, the Homeland Security Geospatial Concept-of-Operations (GeoCONOPS) is a blueprint that identifies best practices, technical issues, and data sources for effective geospatial information and tools to support incident management in the United States. It has been recognized that there is a general lack of standardized processes for geospatial data management in major disasters, leading to the development of the GeoCONOPS concept. This provides sources for appropriate data (in the United States), identification of missions and stakeholders, sources for best practices, and technical capabilities. This is monitored and guided by the geospatial interagency oversight team (GIOT). Version 5.0 of GeoCONOPS was released in June 2013 and is available online at https://government.hsin.gov/sites/GIS. FEMA offers three online courses on this subject through their FEMA Emergency Management Institute at https://training.fema.gov.

Open Source

GIS systems can be expensive, complex, and difficult to learn and use. A major recent development has been the growth and utility of Open Source GIS and image processing systems that provide excellent capability without the cost of commercial software, and which are often smaller in size and easier to learn and use. Open source GIS software, such as QGIS (http://qgis.org), provide complete GIS, GPS,

Fig. 7.2 Open Source QGIS running on an Android tablet (Image courtesy the QGIS project)

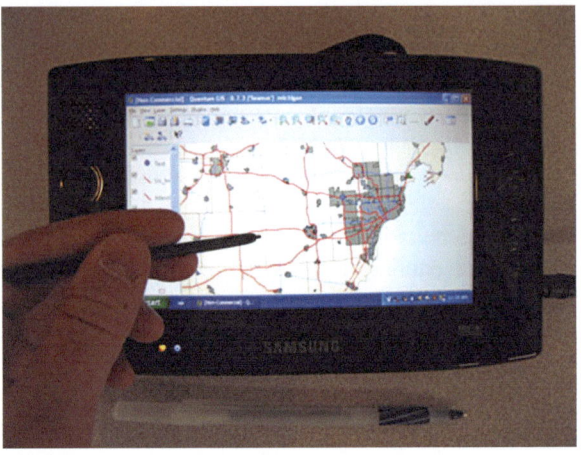

and remote sensing capabilities in several languages that run on Windows, Mac, Linux, and even Android tablets.

One of the benefits of these Open Source systems is that they can actually be loaded onto a flash drive, along with a GIS database, and can run the entire GIS from the flash drive on any computer. There is no need to download the software, have a license, or to compile the code. It will run directly from the flash drive. Having several such 'GIS-on-a-stick' systems pre-loaded allows responders to quickly access important data in the field and begin the response using their own laptops or any available computers in the field (Fig. 7.2).

GIS Systems and Examples in Disaster Management

The QGIS GIS project has recently released the InaSAFE disaster impact modeling system, which was originally conceived by the Indonesian National Disaster Management Agency (BNPB) and Australia's Department of Foreign Affairs and Trade, through the Australia-Indonesia Facility for Disaster Reduction and the World Bank Global Facility for Disaster Reduction and Recovery (World Bank GFDRR). This acts as a plug-in to the Open Source QGIS to provide vital disaster response capabilities by producing realistic natural hazard impact scenarios to support improved and more timely planning, preparedness and response. It allows a simple user interface that integrates data from various sources into likely impacts in disaster events. It is Free and Open Source and can be freely downloaded either as a part of the QGIS or from the inasafe.org website (Fig. 7.3).

The InaSAFE plugin uses existing infrastructure, population, demographics, and hazards data to create models of impacts. Several of the data layers created for the Indonesian prototype were created using OpenStreetMap, as the data did not previously exist. Scenarios for floods, earthquakes, tsunamis, and cyclones of

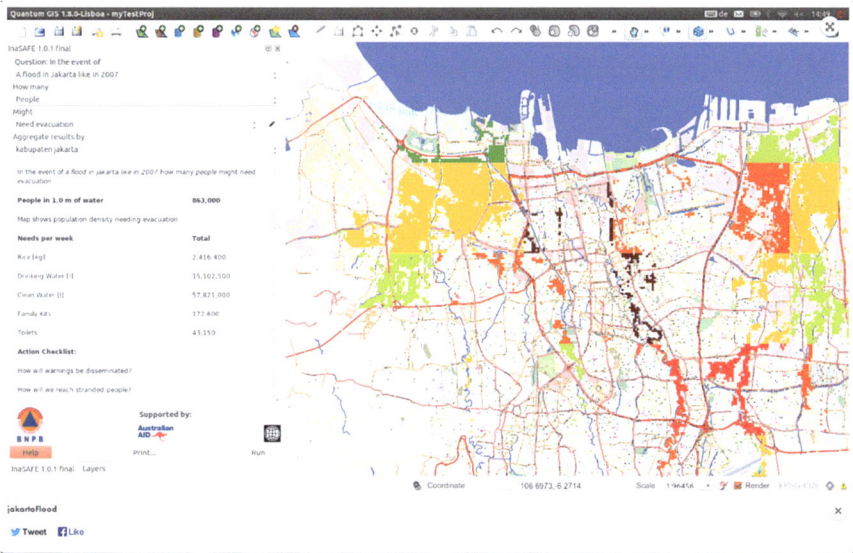

Fig. 7.3 The InaSAFE Open Source disaster modeling plugin QGIS that shows the impact of a flood in Jakarta, Indonesia (Image courtesy of InaSAFE.org)

various intensities are run with building or population demographic data to produce models such as the number of people who might need evacuation, buildings that might be destroyed, etc.

Figure 7.4 shows one example output of the InaSAFE modeling. It shows the total number of people needing evacuation and needed emergency supplies, and the number and type of buildings that will be affected.

U.S. National Weather Service Online Damage Assessment Toolkit

Integrated GIS/GPS systems can now be used for quick damage assessment, and a prototype National Weather Service system, the NWS Damage Assessment Toolkit, has been tested in a variety of disasters, including the Moore, Oklahoma, tornado of 2013. This toolkit integrates data from multiple sources onto laptop computers and hand-held systems, which are integrated into a central database using the Internet, where they are made available to all authorized users through a WWW-based GIS. Various responders, including the Red Cross, local government, other NGOs and police and fire can all access and share data in real time. For the Moore, OK tornado, NWS staff collected over 700 individual GPS points and over 130 ground photos. After quality control, the data were quickly made available to emergency mangers, the media and the public (Fig. 7.5).

Fig. 7.4 Sample InaSAFE output showing the number of people affected, emergency supplies needed, and types of buildings impacted by a flood (Image courtesy of InaSAFE project)

The disparate data from satellite imagery, GPS, GIS, and weather data can now be integrated into real-time, operational alert and warning systems. An example of real-time weather alert GIS, which incorporates many of the systems described above is the TerraMA2 system developed by the Brazilian space organization, INPE. This fully Open Source software system provides an operational system for environmental risk monitoring and real-time alerts, as well as risk mapping and mathematical models to define risk levels, as shown below (Fig. 7.6).

GIS data are integrated with real time and dynamic data from weather stations, in situ measurement systems, ocean buoys, and air quality stations, in addition to weather radar and satellite imagery. These are integrated with various GIS vector risk layers such as rivers, roads, pipelines, and populated areas. Raster data from various satellites are also integrated. Model-based analysis is conducted using defined Boolean logic operations to define levels of alerts. Weather data can be acquired and processed as the various data are received. The output data can consist of maps, web data, and warning and alert data provided by email, RSS, or SMS text to specific geographical areas. The system is currently being used on an operational basis by various Brazilian governmental agencies, including civil defense, public health, and environmental management for a wide range of hazards. Additional information is available at www.dpi.inpe.br/terrama2/english.

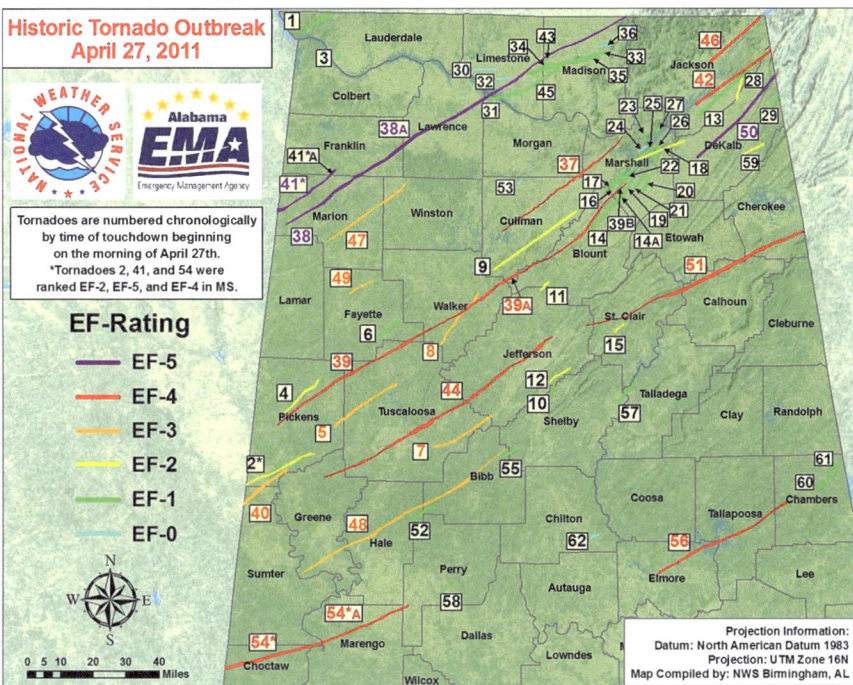

Fig. 7.5 Showing the NWS online damage assessment toolkit of the April 27, 2011, Alabama tornado outbreak (Image courtesy of the National Weather Service)

Data Dissemination Systems

GEONETCAST is an innovative global system for the rapid and broad distribution of disaster imagery and data using existing geostationary satellite systems and small, 1.8-m dishes on the ground. It was established to distribute Group on Earth Observation (GEO) and Global Earth Observation Systems of Systems (GEOSS)-related data and information using a low-cost ground system. GEOSS was established to foster coordinated and sustained observations of Earth using satellites (www.geoss.org). There are three operational components, GEONETCAST America (operated by NOAA), Eumetcast (operated by Europe) and covering Europe and Africa, and FengYungCast (operated by the Chinese Meteorological Agency CMA), and the system allows for sharing and distributing data (Fig. 7.7).

Data distributed includes weather, disasters, climate, air quality and more that is provided on a 24/7 basis. The products on the GEONETCast Americas broadcast can be found at http://www.geonetcastamericas.noaa.gov/docs/gnc-aProductsList033011.pdf.

The system in the GEONETCAST America system, as an example, uses the Intelsat 21 commercial telecommunications satellite and distributes various imagery and data throughout the Americas. The system design is similar to satellite

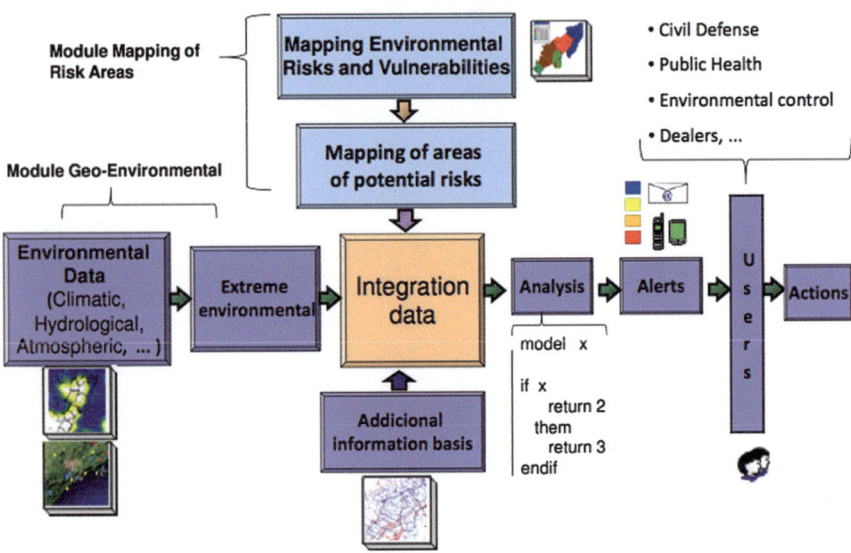

Fig. 7.6 The INPE TerraMA2 system architecture (Image courtesy of INPE of Brazil)

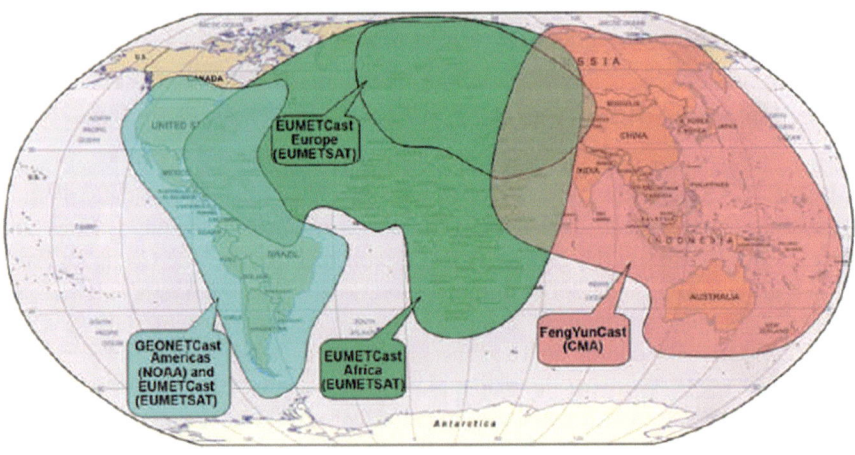

Fig. 7.7 GEONETCAST coverage (Image courtesy of NOAA)

direct broadcast TV but is simpler. A small dish uses a cable to connect to a computer with software and a specialized logic board to transform the images. It can transmit data at a rate of 2 Mbs per second, and there are plans for an upgrade to 12 Mbs per second (Fig. 7.8).

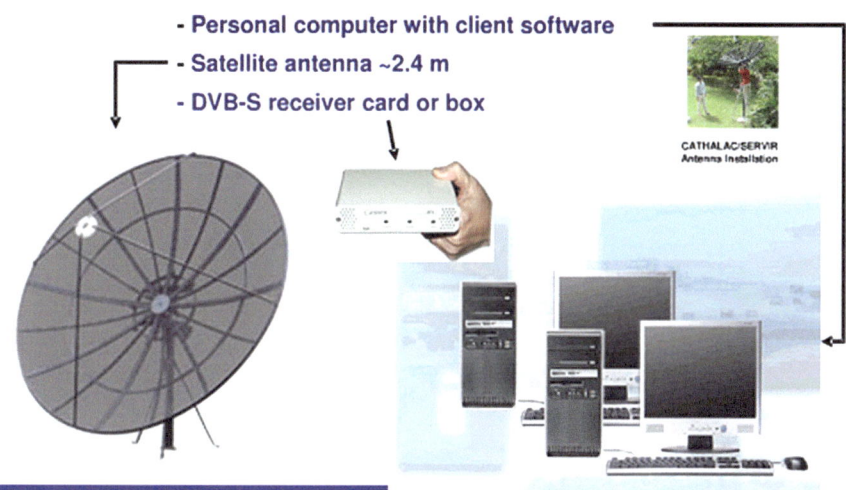

Fig. 7.8 A GEONETCAST ground station configuration (Image courtesy of NOAA)

The system uses a 1.8-m dish on the ground, and installation cost is US$3,000 to $5,000, but this varies due to tax and import duties. The service is free and is broadcast in the clear. Once it is installed there are no additional costs. No Internet connection is needed, which is useful in disasters or remote areas, as the ground system only needs electrical power and can be run for an extended time with a small generator. The system is highly reliable, with over 5 years with no distribution failures. The global GEONETCAST system transfers over 5,000 individual images and files per day. The system requires a line-of-sight view with the satellite, so it does not work in major urban landscapes in between tall buildings.

GEONETCAST is currently operational in Argentina, Brazil, Costa Rica, El Salvador, Mexico, Panama, the United States, and Venezuela, with plans for several more connections in Barbados, Belize, Columbia, Guyana, and Haiti, but with only one or a few systems in each country (Fig. 7.9).

NOAA pays for the operation and satellite transmissions, but INPE of Brazil is now an active partner and it also provides data. GEO weather images are distributed every 30 min, with precipitation, daily forest fires, convective systems, and other images as well as text-based weather forecasts. It has a training channel to improve technical skills of users in meteorology and other topics. In November 2014 there will be new GOES East satellite in operation. It will be able to provide TIFF images in three channels every 30 min with a 4-km resolution. In the case of hurricane conditions, it can be used in a fast-scan mode to provide a localized image every 5 to 10 min.

The GEONETCAST organization is an associate of the disaster charter, and images and data derived through the charter can be distributed as needed through GEONETCAST. High resolution charter satellite images can be transmitted

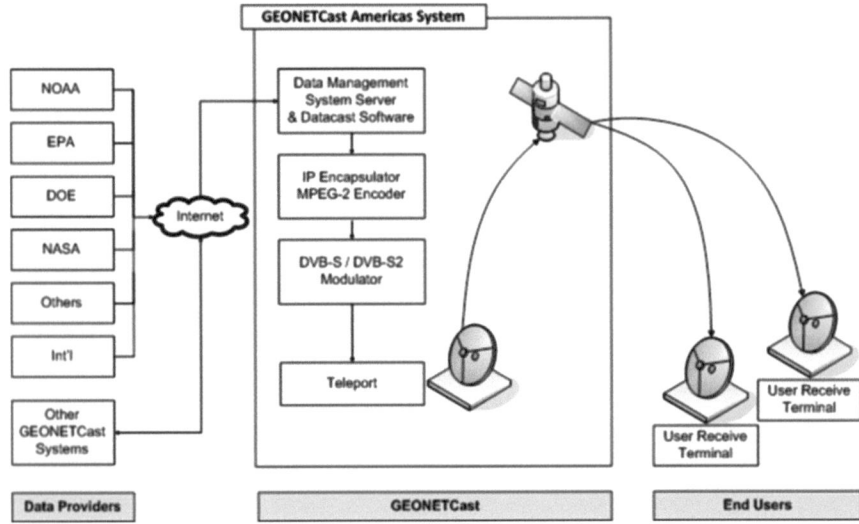

Fig. 7.9 GEONETCAST architecture (Image courtesy of NOAA)

quickly over wide areas, which can significantly improve disaster response based on the most recent imagery information, as well as weather forecasts and other information.

Early Warning Systems

The Global Disaster Alert and Coordination System (GDACS) was created in 2004 by the European Commission Office for Humanitarian Aid (ECHO), U.N. Office for Coordination of Humanitarian Affairs (OCHA), and the U.N. Operational Satellite Applications program (UNOSAT). This system is designed to increase the use of reliable and accurate disaster alert, and to foster improved international cooperation in major disasters. Initially funded by the European Commission, a web-based service was established that integrates disaster warning information. Automatic information exchange between the various participants is possible, and many governments and disaster response organizations routinely use the Virtual On Site Operations Coordination Center (VOSOCC) and ReliefWeb in major disaster responses. Over 10,000 subscribers currently are registered to make use of the system (http://www.gdacs.org) (Fig. 7.10).

The GDACS system includes global monitoring for major disasters, including earthquakes, tsunamis, floods, volcanic eruptions, and tropical cyclones. It approaches this through a system of systems approach using Open Source tools to promote interoperability. It uses the common SML format really simple syndication

Early Warning Systems

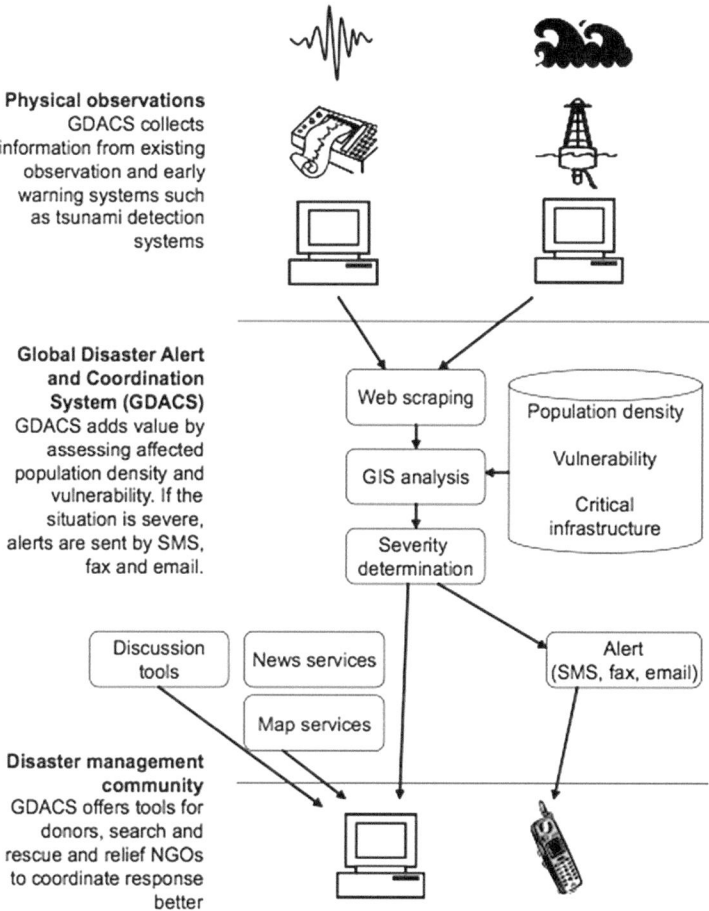

Fig. 7.10 Chart showing the structure of the GDACS system (Image courtesy of OCHA)

(RSS) for exchange of messages, so that nearly any organization or user can have access.

A second aspect of the system is the use of the GLIDE (Global Identifier for Disaster) format, which was created by the Asian Disaster Reduction Center, which is a standardized identifier format to allow multiple organizations to link their databases. This format includes a two-letter identifier of the type of disaster (EQ for earthquake), the year, a six-digit sequential disaster number, and finally the three-digit ISO country code for the location. This allows quick sharing of information across different systems and language. The GDACS secretariat is located at OCHA in Geneva.

Chapter 8
Major International and Regional Players

The United Nations

The United Nations is one of the most important players in disaster relief and in the promotion of space technologies to aid in disaster preparation, response, recovery and mitigation. Following are some of the agencies within the United Nations that provide disaster relief services.

The U.N. Space-Based Platform for Information for Disaster Management and Emergency Response (UN-SPIDER)

The U.N. General Assembly and the Committee on the Peaceful Uses of Outer Space (COPUOS) have recognized the importance of space data for disaster response and sustainable development and has mandated the creation of appropriate programs to respond to these needs. Several programs have been created to cover all areas of space within the overall U.N. umbrella. One of these is the UN-SPIDER, which was created in 2006 by the U.N. General Assembly, to provide satellite technology support specifically for disasters. This is done through several focuses, including training courses, simulations, and advisory support missions to build local capacity. The SPIDER web portal is an excellent source of up-to-date information on all aspects of these topics, with over 300 articles, training materials, e-books and more.

The SPIDER Space Application Matrix (http://www.un-spider.org/space-application/space-application-guides) provides a simple, web-based tool on 'Geoinformation for Disaster and Risk Management – Examples and Best Practices.' This document is an excellent compilation of case studies and can be downloaded from their web portal.

The Value of Geoinformation for Disaster Risk Management (VALID) report was published in 2013 and provides an excellent overview of the benefits and technologies in this field (http://www.un-spider.org/about/publication/value-geoinformation-disaster-and-risk-management-valid).

The U.N. Institute for Training and Research (UNITAR)

This is the main vehicle for the U.N.'s operational Satellite Applications Program, which delivers satellite imagery and analysis to both U.N. and non-U.N. humanitarian organizations when disaster strikes.

The International Federation of the Red Cross Red Crescent Societies (RCRC)

This is the world's largest humanitarian network and is one of the major international disaster players and has over 100 million members and volunteers.

There are three parts to the RCRC movement: 187 national societies, the International Federation of Red Cross and Red Crescent Societies, and the International Committee of the Red Cross. These, together, operate around the world to prepare for and respond to disasters. The RCRC movement operates in accordance with the fundamental principles of humanity, impartiality, neutrality, independence, voluntary service, unity, and universality. The national societies play a unique role, providing disaster relief, collecting blood, supplying shelter and food, and supporting health and social programs.

The Committee on Earth Observation Systems (CEOS)

This group is dedicated to the idea of advancing satellite data democracy. It was founded in 1984 and seeks to coordinate space-based Earth observation activities. CEOS has 30 members and 22 associates. Participating agencies work together to address critical scientific questions and to harmonize satellite mission planning to address gaps and overlaps. It specifically focuses only on space assets, unlike GEO, which also includes in-situ systems as well. Disaster societal benefit is one of their main areas of focus. The idea of data democracy was started by the Council of Scientific and Industrial Research (CSIR) in South Africa. Data democracy promotes open access, capacity development, software tools, and data dissemination. Working to make Shuttle and Radar Tomography Mission (SRTM) data available especially to the underserved nations in the developing regions of the world was one of their initial goals.

This committee also produces and updates the Earth Observation Handbook (http://www.eohandbook.com), which is the most comprehensive and up-to-date document on the status and capabilities of the world's Earth observation systems and operators.

The CEOS Visualization Environment (COVE)

COVE is a web-enabled capability that allows users to identify what data are available and uses Google Earth to display the coverage of various satellite instruments (www.ceos-cove.org). There are currently 109 satellites in use for Earth observation and over 50 can be tracked using the online COVE system. In 2014 CEOS announced a new disaster response management strategy, specifically designed for the needs of the disaster community. This will cover flood areas, seismic events, and volcanic eruptions.

The Group on Earth Observation (GEO) and the Group on Earth Observations System of Systems GEOSS

These are two important global organizations working in this field. GEO was created in 2005 to create a coordinated and sustained system of global Earth observation (EO) systems. It currently has 90 member states, which participate on a non-binding and voluntary basis to share and coordinate Earth observation activities. There are also 67 participating organizations, such as the United Nation, and many other international and regional participants. The idea is to not create a new or single international satellite program but to coordinate the world's many EO systems better. It also includes coordination and use of important in-situ components. These include resources around the world such as ocean buoys, rain gauges, etc.

GEO has an important data sharing mission, which is to seek open and full exchange of data for all users with minimum time delay and cost, but this mission is non-binding. There are both top-down national and bottom-up approaches, where the scientists collaborate and push governmental entities to be involved.

GEOSS's primary focus is, as its name indicates, to be a system of systems. Its mission is to integrate and interconnect all of the established and new remote sensing systems in a practical way and support informed decisions for all aspects of society, but not to define or operate specific systems.

Issues that are addressed include the uncertainty of continuity in individual systems, large temporal and spatial gaps in data sets, limited access to data sets, and the lack of mechanisms for interoperability. GEOSS has three main areas of effort. These are the coordination of the building blocks for Earth observation, achieving societal benefits, and cross-cutting systems. The architectural aspects of GEOSS (i.e., the building blocks) are to address the different ways that data are archived and accessed and to promote standardization so that data can be accessed through any portal.

Currently there are more than 7 million resources, including image catalogs, and over 65 million individual 'granules' of data, such as individual images, rain gauge records, etc., available via the GEO web portal. The difference between Google and the GEO web portal is that Google is only for locating and accessing remote sensing and related data.

The capacity building GEO effort is designed to increase use of EO and derived products. There are nine societal benefit areas, and disaster response is one of the nine included, but all others are also relevant, such as climate, agriculture, and health. The disaster activity area is working to improve disaster risk management and reduction.

The GEO web portal has established several permanent geohazard supersites, areas with particular vulnerability, such as the North Anatolian and San Andreas faults, where frequent hi-res SAR imagery monitors information along with in-situ systems. GEO, CEOS, and NASA are working together on a disaster sensor web concept that will allow tasking sensors and have an integrated approach.

FEWS Net, the Famine Early Warning Network III

The goal of FEWS Net (http://earlywarning.cr.usgs.gov/fews) is to provide early warning of food problems to improve the systems related to famine, food markets, weather conditions, and crops in countries. The community of 22 nations, including USGS, NOAA, USDA, and NASA participate. The system uses a mix of information, including field verification, market prices for commodities, rain totals, and crop growth measurements by satellite to predict famine. They compute NDVI from moderate resolution satellites to monitor crop development and generate patterns of rain vs specific growing seasons using TRMM. The system provides very accurate data. Every 3 months a new set of predictions are made, and the last predictions are analyzed.

Asia Disaster Reduction Center (ADRC)

This was established in Kobe, Japan, in 1998 after the great earthquake in that region. The goal of the ADRC is to enhance disaster resilience and sustainable development within its member states throughout Asia. During a disaster event in the region, the center releases data and information and provides technical assistance.

Sentinal Asia

This is another important resource. Asia suffers disproportionately from global disasters, so Asian nations are very active in the use of space technologies to aid with them. RADAR data are particularly useful in cloud-covered regions. This program began in 2006 and consists of 8 international organizations and 51 participating entities from 20 Asian nations, and aims to promote international cooperation to monitor disasters in the Asian-Pacific region. The Japanese space agency, JAXA, provides satellite data and technical assistance in cooperation with the United Nations and others and operates a facility in Bangkok, Thailand, for the organization. Satellites from India and Thailand are also involved. Requests for data are placed through the Asian Disaster Reduction Center, and the data are acquired and posted on program websites.

The Pacific Disaster Center

This was created by the U.S. government and is located in Maui, Hawaii. It is yet another resource that provides a web portal that can provide valuable space-related data to aid in disaster preparation, response, recovery, and mitigation. It has also generated computer simulations, using space and local data to show the dangers of construction in areas prone to flooding and severe tsunami impact. To participate in PDC programs and utilize key space data from the web portal one must officially register to obtain access.

Chapter 9
The Emerging World of Crowd Sourcing, Social Media, Citizen Science, and Remote Support Operations in Disasters

The concepts of citizen science and crowd sourcing are new to the disaster community but have great potential. Crowd sourcing is more focused on non-skilled assistance, while citizen science is more focused on using skilled workers in a remote support mode to provide specific, remote sensing GIS and mapping skills without being physically deployed into the disaster area.

Crowd Sourcing

Crowd sourcing is the practice of getting some needed service, idea, or work by soliciting contributions from a large group of people, usually using the Internet and online communities, rather than from traditional work sources. Often this work is voluntary and can be used for distributed work, problem solving, and other uses. The Haitian earthquake in 2010 provided a fascinating study in both the benefits and problems of crowd sourcing for disasters. This was the first time a large number of young, computer-savvy students decided to apply 'their' social media and crowd sourcing skills. Many self-deployed to Haiti and others worked remotely from the United States and around the world. The American Red Cross also raised over US $32 million for their Haitian relief with a simple SMS text donation system. There was very little spatial data available for Haiti and the capital of Port-au-Prince, and a spontaneous process of mapping the affected area using OpenStreetMap produced an extraordinary response and provided a base map that was used throughout the international disaster response community. Over 600 people contributed to this effort around the world, with most working on their own computers in the United States. Many were young students working in coffee shops in an ad hoc network (Fig. 9.1).

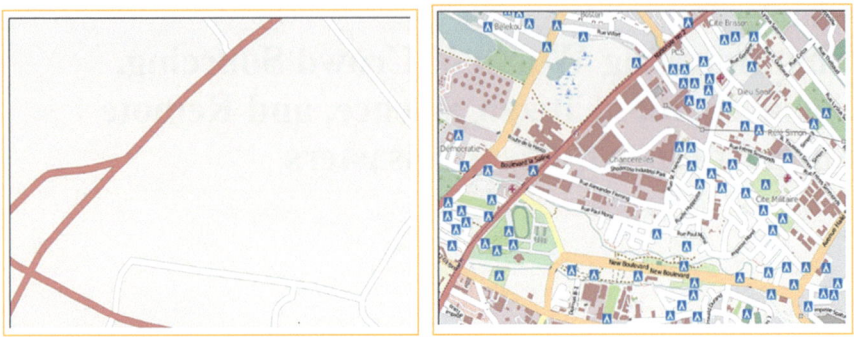

Fig. 9.1 OpenStreetMap for Port-au-Prince, Haiti, before and after the earthquake (Image courtesy OpenStreetMap.org)

Social Media

The emerging world of social media is starting to have an impact on the disaster management community. In the Haitian earthquake of 2010, the cell phone system was surprisingly operational after the event, and Project 4636 was set up to allow people with cell phones to use short messaging services (SMS) to call the number 4636 for free. At its peak, the system was receiving over 5,000 calls in 1 h. An ad-hoc system was set up that consisted of someone sending an SMS text to the 4636 shortcode. This was then routed onto the Emergency Information Service (EIS), where it was tracked and then forwarded to the www.crowdflower.com website, where a Haitian volunteer or other person logged it into the system and translated the message to English, adding appropriate data such as geographical information. The information was then turned into a standardized report that was distributed to appropriate response organizations, such as Urban Search and Rescue (USAR) teams or the Red Cross, for action in the field (Figs. 9.2 and 9.3).

This innovative use of SMS, crowd source mapping, and other new technologies has promoted an active debate in the emergency management community. How can the disaster response community better incorporate these new tools? Was Haiti the new standard, or was it a unique confluence? The document entitled Disaster Relief 2.0 (http://www.unfoundation.org/assets/pdf/disaster-relief-20-report.pdf) presents a very positive case for the former, but there were also significant problems, including false reports. A U.S. Congressional Research Service report of 2011 (http://www.fas.org/sgp/crs/homesec/R41987.pdf) takes a more cautious view. This report notes the potential unknown cost to governments and the problems that can occur if incorrect information and intentionally false information is provided. The report states "While there may be some potential advantages to using social media for emergencies and disasters, there may also be some potential policy issues and drawbacks associated with its use."

A survey of U.S. governmental agencies showed that 9 out of 10 disaster response agencies are not staffed and have little or no budget to support social

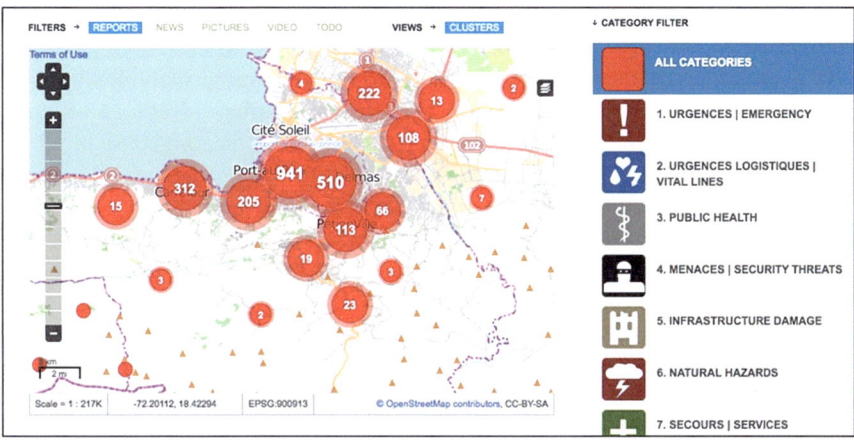

Fig. 9.2 Map of Project 4636 and volume of SMS messages in the Haitian earthquake (Image courtesy of www.crowdflower.com)

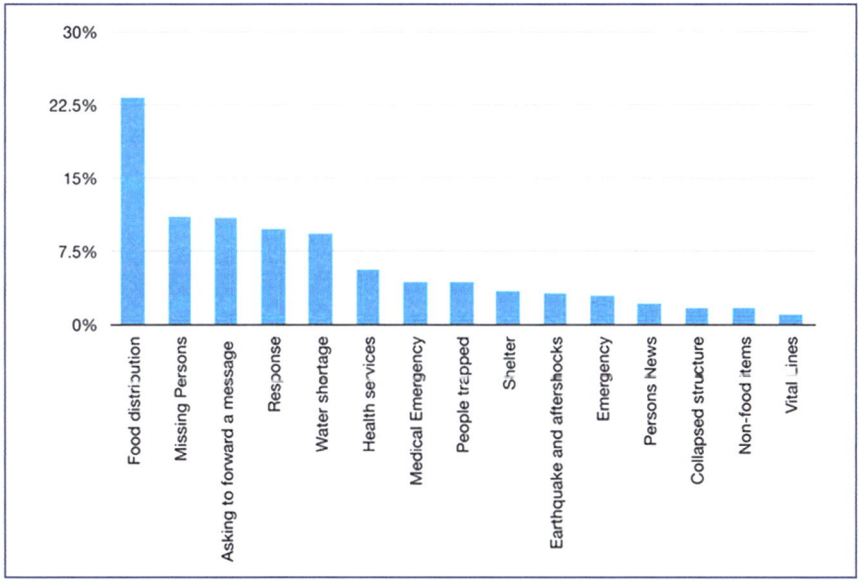

Fig. 9.3 Graph showing SMS text requests for assistance in the 2010 Haitian earthquake (Graph courtesy of Project 4636)

media activities. During Hurricane Sandy in 2010 there was an incident of an individual who was purposefully sending false information to disaster responders. Issues of the validity of data, the sheer quantity of data and the need for filtering, and the large and changing number of volunteer organizations remains an issue.

Such concerns as the potential liabilities and the ownership of such data have yet to be addressed.

The response of those involved in these crowd sourcing activities to these concerns was to create the Digital Humanitarian Network (http://digitalhumanitarians.com). This is a network-of-networks clearinghouse of all the various players in this voluntary and technical community (V&TC), which has been created to provide for a single point of contact between all such volunteers and the official disaster response organization. The DHN process resembles the International Disaster Charter, in that a request is made to activate the network, and coordinators from the Humanitarian OpenStreetMap team, MapAction, CrisisMappers, and the GISCorps review the request. The process mirrors the activation process of the International Charter for Disaster Management, but is different in that any organization or group can request assistance. If activation is accepted, the coordinators create a 'solution team,' which then manages the response and provides a single point-of-contact to the humanitarian responders for all of the V&TC group activities. The DHN can provide a wide range of assistance, including geo-location of event and infrastructure data, creating maps, monitoring social media, GIS data creation and analysis, remote sensing data analysis, and more (Fig. 9.4).

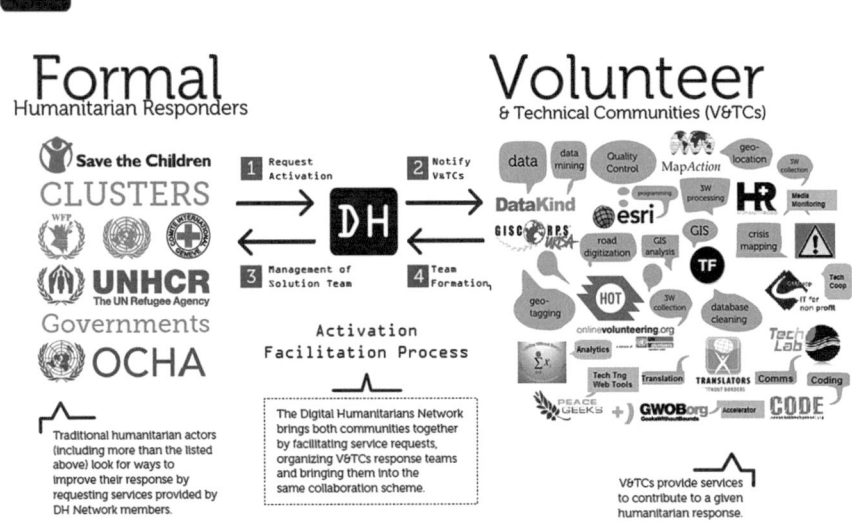

Fig. 9.4 Formal and volunteer responders to global disasters (Graphic courtesy of www.digitalhumanitarians.com)

Citizen Science and Remote Support Operations

Citizen science is another aspect of this community where trained professionals, as opposed to well-intentioned volunteers, provide valuable 'back office' support to the disaster response using their own computers, software, and Internet around the world. There are several organizations that focus on this type of effort, including the URISA GISCorps. This is a group of professional GIS and remote-sensing specialists who volunteer, again using their own computers, software, and Internet resources, to process data without deploying to the disaster area.

This type of remote support operations is very valuable in disasters and has been used successfully several times. One benefit of this approach is that multiple individuals can work on small tiles of data, which can then be aggregated to cover large areas. Also, workers around the world can process data in a continuous flow, and as one region ends work for the day, the next group in the next time zones can pick up the work and then pass it on in turn. This can drastically increase the speed of large data analysis in the vital early hours of the disaster response. This technique has been successfully used in flooding in Australia, the typhoon in the Philippines, and other locations.

Major Players in the V&TC Community

These now include Crisis Mappers, The Humanitarian OpenStreetMap team (HOT), Digital Humanitarian Network, GISCorps, and others. The rapidly evolving world of these many new tools will have a major impact of disaster management and response as we move into the future.

Chapter 10
International Treaties, Non-binding Agreements, and Policy and Legal Issues

The International Charter on Space and Disasters

In 1999, growing out of the UNISAPCE III conference in Vienna, the International Charter on Space and Major Disasters was created to institutionalize and structure a process whereby satellite remote sensing imagery and data can be provided by national and regional space agencies to major disasters in a coordinated manner (http://www.disasterscharter.org). The charter was originally started by ESA and the French space agency CNES in 1999, with Canada signing in 2000 and the United States (NOAA), the Argentine space agency, USGS, JAXA, GNSC, China National Space Administration, DLR, KARI, Eumetsat, and IMPE later signing as well (Fig. 10.1).

The purpose of the charter is to provide a standardized and practical process for sharing satellite imagery from the various national space agencies and satellite operators to disaster responders in a timely manner. Each member agency agrees to provide data and assist in major disaster operations (Fig. 10.2).

There are several formal processes for charter activation. Originally, only a dedicated representative of a cooperating nation could institute the process and request activation. In 2010 the charter was extended to provide open universal access. This means that now any national disaster management authority can submit requests for activation. Thus the affected area will not have to be a charter member state as before, but will still have to register a point-of-contact to initiate the process.

Figure 10.3 shows the sequence for the International Disaster Charter activation. An authorized user calls or emails a 24-h duty operator to start the process. This duty operator checks the identity of the requester and verifies that the User Request Form has been properly completed. Then the information is sent to the emergency on-call offices. The request is then reviewed, and the process of preparing the satellite data acquisition plan is started. This plan seeks to use the correct mix of appropriate and available resources. Imagery comes from a wide range of sources, including governmental and commercial satellites. The data are acquired and

Fig. 10.1 Official seal of the International Charter on Space and Major Disasters (Graphic courtesy of the International Disaster Charter: www.disastercharter.org)

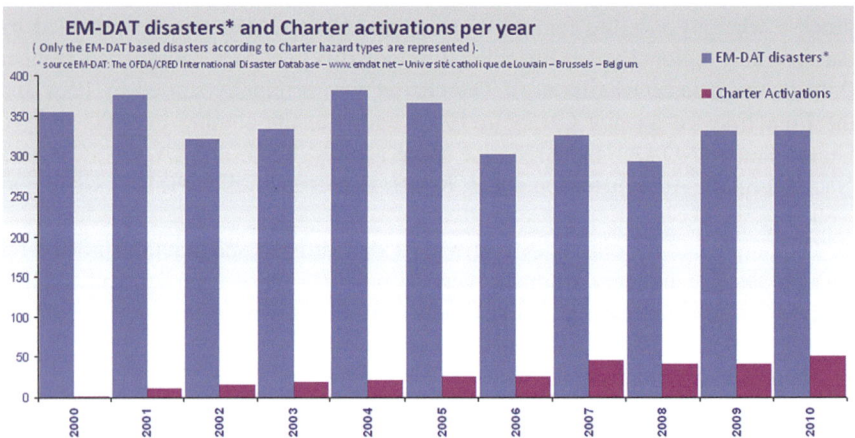

Fig. 10.2 Graph showing the International Disaster Charter activations from 2000 to 2010 (Graphic courtesy International Disaster Charter: www.disastercharter.org)

delivered, and a project manager is appointed to oversee the process. Data are now routinely requested, acquired, processed, and delivered to the requesting entity in a maximum of a few days and often sooner. The data is then provided as required through to the end of the disaster. All data are acquired, processed, and provided free of charge.

The charter has now been activated in response to over 400 major disasters in 120 nations over a 15-year period. All data are provided free of charge for use by the disaster response, but there are limitations of the follow-on use of some of the imagery after the disaster response is completed. General inquiries can be directed

The International Charter on Space and Disasters

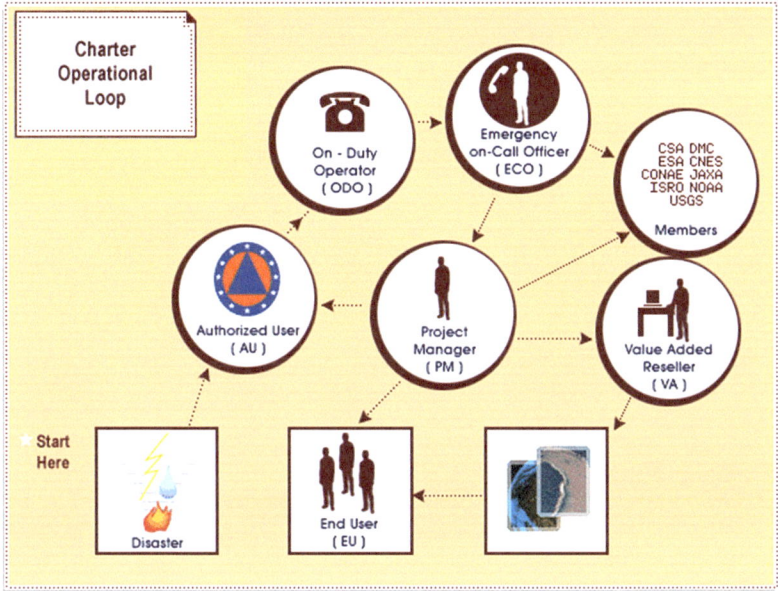

Fig. 10.3 Chart showing the process for activation of the charter (Image courtesy of disastercharter.org)

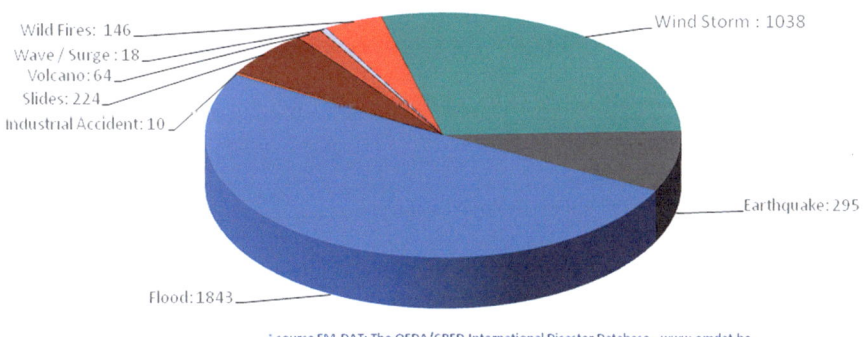

Fig. 10.4 Distribution of major disasters 2000–2010 (Graphic courtesy of OFDA/CRED)

to webmaster@disastercharter.org. It is very important that disaster managers know who the authorized users for their country are, and understand the process for activating the charter, what data are available, and how these can be used (Figs. 10.4, 10.5, and 10.6).

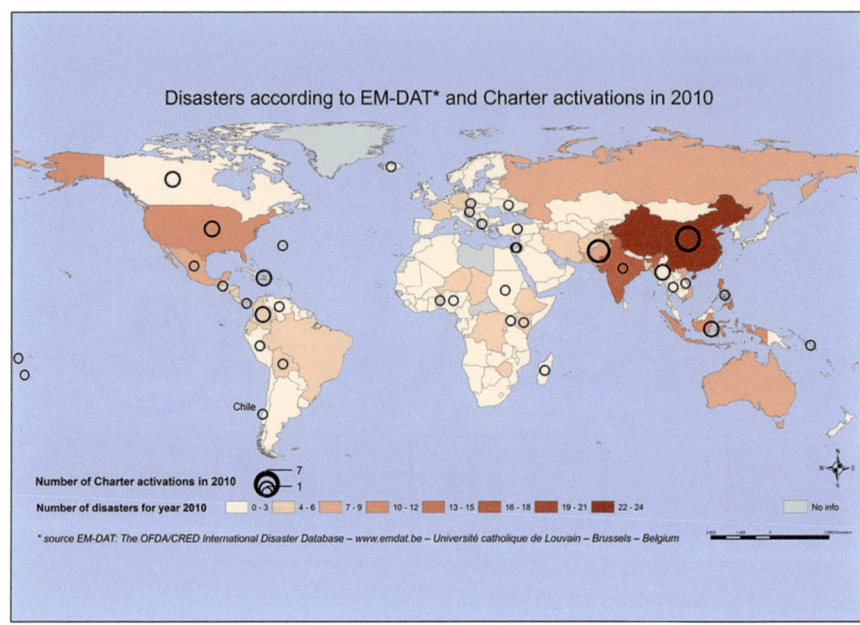

Fig. 10.5 Map showing the location and number of charter activations in the year 2010 (Graphic courtesy of disastercharter.org)

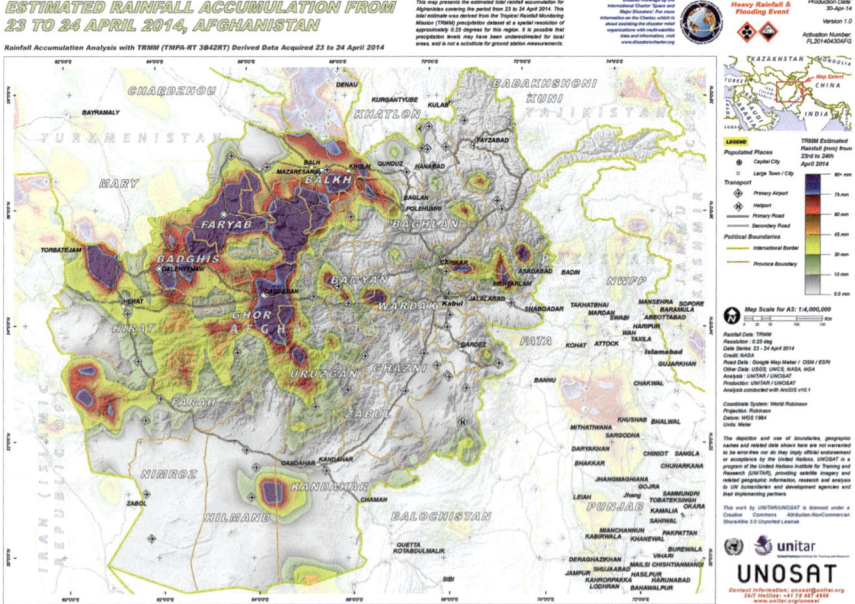

Fig. 10.6 A NASA map showing rainfall accumulation for Afghanistan April 23–24, 2014, derived from the TRMM satellite. This is a typical product from the Disaster Charter (Image courtesy of NASA)

The Tampere Convention on the Provision of Telecommunication Resources for Disaster Mitigation and Relief Operations

This is an international treaty originating in 1998, coming into force in 2005, that provides for satellite communications equipment to be easily shipped across borders to support disaster needs without customs duties or delays. It calls on states to promote the use of telecommunications in disaster response, and has been ratified by 46 states so far, describing procedures for requesting and providing telecommunications resources across international borders and the waiver of regulatory barriers such as frequency allocation, importation of equipment, and the movement of relief teams across national borders in the event of a disaster. Although this treaty is in effect, it is little known and delays still occur. The development of satellite telephones has also reduced the need for larger systems, although they still play a role.

International Non-binding Agreements

There are several important non-binding agreements between states that play an important role in international disaster response. The main one of these is the U.N. General Assembly resolution 46/182, which was unanimously adopted in 1991. This defines the role of the United Nations in assuming a coordinating role in international humanitarian situations when a national government requests external aid and assistance. The resolution specifies U.N. mechanisms to create and promote coordinated and effective international humanitarian activities. The most important of these is U.N. General Assembly resolution 46/182. This defines the role of the United Nations and its agencies in coordinating international humanitarian relief when any member state requests its support. Mechanisms are in place to establish U.N. coordination through the Central Emergency Response Fund (CERF), the Consolidated Appeal Process (CAP), the Emergency Relief Coordinator (ERC), and the Inter-Agency Standing Committee (IASC), which is most appropriate to the case in question.

Legal and International Policy Issues

There are several very important policy and legal issues relating to remote sensing, GNSS, GIS, and new, emerging technologies such as crowd sourcing and community remote sensing for disaster management.

The legal issues involving acquiring immediate visas for relief workers, work permits, recognition of professional qualifications, customs duties and import regulations, civil liability for relief staff, aircraft over flight and landing rights, NGO registration, and more. The use of unmanned drone aircraft in disaster zones is another complex issue that is still being sorted out.

One constant is that policies and laws for spatial data are confusing and constantly changing. There are complex issues of data privacy, data liability, national security, intellectual property rights, and economic policies. In almost all instances the necessary laws and policies are well behind the current state of technology. Countries have varied laws and are sensitive about the collection of information by individuals and non-governmental organizations. Such functions are traditionally within the realm of national security agencies and national governmental agencies, and these traditional limitations are difficult to overcome.

Issues of privacy concerns vary culturally and nationally. It is now pretty much a given that technology always precedes the relevant law and policy and there is always catching up to do. There are recent examples, such as errors in car navigation GPS systems. In the state of Georgia in the United States, a contractor destroyed the wrong house because they used the GPS system to identify the address, and destroyed the house next door to the intended structure. Who owns all this data created by crowd sourcing or data mash-ups? Who is liable for spatial data that is aggregated via crowd sourcing? Who has access and who decides when the data should be updated or retired once it is out of date? What restrictions can be placed on all this both within and between nation states? The world of spatial data supply and access is changing very fast, and there are many disruptive technologies that are at work and this will only accelerate. Keeping up with all the various issues and changes will be difficult, but this is a very important aspect of the intersection between advanced spatial technologies and disaster management.

Chapter 11
Future Directions and the Top Ten Things to Know About Space Systems and Disasters

Future Directions

Space technologies provide vital information to all phases of the disaster cycle and form a growing suite of tools that effectively support the disaster management community worldwide. What emergency managers need are reliable tools and proven partners that fit into their existing organizational structures and work flows. These two vital capabilities can expand their ability to respond effectively. They need actionable intelligence, not more and more complex computer systems and mountains of data. They also need tools that are reliable, robust, and that they can incorporate into their training and response structures. Ad hoc and experimental advanced technologies that are not reliable or consistently available are of little interest and can have a negative overall effect because they can make the EM world wary of adopting useful advanced technologies. It is vital that the space and IT communities study and learn the realities of the emergency management communities so that we can meet their needs with appropriate tools and capabilities.

The emergency management (EM) systems on one hand and the space/high technology networks on the other are two very different cultures. We have seen that emergency managers work in a culture that is highly structured, with a hierarchical command and control organization that often mirrors a military structure and which does not readily incorporate untried innovation. Lives and property are at stake in a very high visibility environment. In this it is similar to the human spaceflight and large space agency cultures. It is common for new, high tech, and expensive 'magic bullets' to be promoted to the EM community (for a price), and it has been burned in the past with expensive and difficult to use new tools that did not perform as needed and were difficult for working EMs to learn and use under difficult field conditions.

The academic and research communities that generate such new innovations come from a very different set of cultures that are, in many aspects, opposites. The academic research and scientific communities are unstructured, not hierarchical, and are comfortable with a looser organizational structure. There are less defined leadership roles, and they are quick to adopt new technologies and approaches. Some 'high tech' proponents of the use of social media in disaster response have recently made significant efforts to incorporate these new capabilities of social media and community participation into the EM response, but with varying degrees of success. Loosely organized volunteer organizations are more comfortable with a bottom-up and distributed structure with less defined leadership roles. These volunteers are trying hard to get the EM community to incorporate their ideas, but there is a real culture clash and also a generation gap between these worlds.

Ultimately, the new generation of volunteer high tech relief groups are today not responsible in the same way as the professional EM community, which bears the responsibility for lives and property in its jurisdictions. Volunteers can choose to deploy or not for a given incident, while the professional EM community bears the ultimate responsibility for the success or failures of the response, and ultimately for lives and property. It is a complex and difficult problem, but one that can be surmounted.

In the Summary of a Workshop on Using Information Technology to Enhance Disaster Management (National Academy of Sciences, 2005), it was noted that existing IT (information technology) 'push from the top' as opposed to 'pull from the EM community' is not working, and that the rate of technology innovation and change is greater than public safety and EM organizations can realistically adopt. Training responders in new technologies remains difficult, and it is clear that IT and space technologies must be in routine use by the EM community to be effective, as people fall back on what they know in a disaster. It was also noted that technology that is not included in standard operating procedures and training scenarios will often not be used correctly, if at all, in a real disaster. The report concluded that revolutionary leadership is needed to overcome the political, financial, and organizational challenges to incorporating advanced technologies in disaster management. How is this to happen? Who will provide this? It is simply a matter of time of transition as a new generation of workers enter the EM community.

The space community has provided important capabilities to the EM community, and space assets are now regularly being utilized in the overall disaster response around the world with varying degrees of success. This use will only increase in the future. The space community, with its many capabilities and extensive expertise in advanced technology and large organization management, can provide important leadership in addressing these issues. The use of advanced tools, and new capabilities such as social media, will only increase over time as new and more capable space technologies are developed. The space technology world still needs to do a better job of understanding the EM world, perspective, and organizational structures. It clearly needs to incorporate its capabilities into the national and international EM response systems more effectively. Too often, the 'technology push' approach is used. It is far better to engender an 'end user pull' in seeking to incorporate capabilities. This can occur without the EM community fully understanding the specifics of how new space and IT systems and data can be most effectively accessed, best learned, deployed, and utilized under difficult field conditions. Space technology organizations, agencies, researchers, and advocates could do a better job in this important work if they would take more time and effort to better understand the reality of how the EM world works, including becoming more directly involved in disaster management operations themselves. We need to work harder to fit *our* tools into *their* world in order to be effective. The young, technical, and volunteer communities need to learn to work together better. They also must learn to work with the EM community, and understanding the EM world and perspective is an important aspect of this. In the end, it is too important for us to not succeed in making space technologies as useful as they can be in this vital domain, and technology by itself will not solve this cultural problem. This will only be achieved by working together in good faith, and by seeking to understand the EM culture and their needs and perspectives more fully, and by integrating geospatial data into their daily routines. By working together with them through all aspects of the disaster cycle, we can ensure that these powerful capabilities are fully integrated into their training and response toolkits. They, and the people they serve, deserve no less.

Top Ten Things to Know About Space and Disasters

There are a lot of space technologies and systems that have now been deployed to cope with various types of disasters – both natural and human-made events. It is perhaps easy to be caught up in the details of the technologies, but essentially there are only a few key elements to remember that distill the essence of the tools from space that allow us to cope with disasters in a more effective way around the world. The following ten points represent these crucial concepts.

1. Space plays an important and growing role in disaster management.
2. Space technologies are of growing importance in all four parts of the disaster cycle of preparation, response, recovery, and mitigation.
3. Space technologies are relevant to all nations in disaster management regardless of their level of economic or technological development.
4. A day without space in a disaster would be a bad day on the ground.
5. The integration of many different types and sources of data, using geomatics, is the future. This includes new social media information.
6. You can drown in data, too; and how we filter, distill, analyze, and share the data is important.
7. Social media, citizen science, and crowd sourcing will be more important.
8. The technology community needs to learn the needs and processes of the emergency management community.
9. Sometimes advanced technology is not the answer in a disaster.
10. Disaster can strike you and your family at any time. Everyone needs to be prepared, have a disaster kit, and become informed.

Appendix A: Key Terms and Acronyms

ADPC Asian Disaster Preparedness Center
ANSI American National Standards Institute
API Application Programmer Interface
AO Area of operations
APT Automatic picture transmission from NOAA weather satellites
ARC American Red Cross
Bent Spear A military nuclear incident
Broken Arrow A military nuclear accident
BSS Broadcast satellite service
CAD Computer-assisted design
CAMEO Computer aided management of emergency operations
CAP Civil air patrol
CBRNE Chemical, biological, radiological, nuclear, explosive
CD Compact disc
CDC Centers for Disease Control
CDERA Caribbean Disaster Emergency Response Agency
CDRG Catastrophic Disaster Response Group
CERT Community Emergency Response Team
CI/KR Critical infrastructure/key resource
CIP Critical infrastructure plan
CIPI Critical infrastructure protection initiative
COG Continuity of government, also council of governments
COOP Continuity of operations
COP Common operating picture
COTS Commercial off the shelf
CSIR Council of Scientific and Industrial Research
DA Disaster Assessment
DCO Defense coordinating officer
DDA Detailed disaster assessment
DFO Disaster field office

DHS Department of Homeland Security
DMC The disaster monitoring constellation of satellites
DMORT The Disaster Mortuary Response Team
DOCSIS Digital Over Cable System Interface Standard
DRO Disaster recovery operations (American Red Cross)
DVB-RCS Digital video broadcast with return channel service
EADRCC Euro Atlantic Disaster Response Coordination Center (NATO)
ECRV American Red Cross emergency communication response vehicles
EDXL Emergency Distribution Exchange Language
EEI Elements of essential information
EM Emergency management
EMAP Emergency Management Accreditation Program
EMDC Emergency Mapping and Data Center
EMI FEMA Emergency Management Institute
EMS Emergency medical service
EOC Emergency Operations Center
EOP Emergency operations plan
EPA Environmental Protection Agency
EROS Earth Resources Observation and Science
ERRO Emergency Response and Recovery Office
ESF Emergency support function
ESF #1 Transportation emergency support function
ESF #2 Communications emergency support function
ESF #3 Public works and engineering emergency support function
ESF #4 Firefighting emergency support function
ESF #5 Information and planning emergency support function
ESF #6 Mass care emergency support function
ESF #7 Resource support emergency support function
ESF #8 Health and medical services emergency support function
ESF #9 Search and rescue emergency support function
ESF #10 Hazardous materials emergency support function
ESF #11 Food and water emergency support function
ESF #12 Energy emergency support function
ESRI Environmental Systems Research Institute
FAA Federal Aviation Administration
FCO Federal coordinating officer
FECC Federal emergency communications coordinator
FEMA Federal emergency management agency
FeX Field exercise
FGDC Federal Geographic Data Committee
FIRMS The Fire Information for Resource Management System of NOAA
FSS Fixed Satellite Service
GDACS The Global Disaster Alert and Coordination System of the United Nations

Appendix A: Key Terms and Acronyms

GDIN   The Global Disaster Information Network
GEO   Geostationary orbit
GEOCONOPS   Geospatial concept-of-operations at the Dept. of Homeland Security
GEONETCAST   Geostationary image and weather data transmission system
GEOSS   Group on Earth Observation System of Systems
Geostationary orbit   A circular satellite orbit over the equator at 35,786 km whose position in the sky remains the same to a fixed position on the Earth
GFDRR   Global Facility for Disaster Reduction and Recovery of the World Bank
GIOT   Geospatial Interagency Oversight Team at FEMA
GIS   Geographic information system
GISMO   Geographic Information Systems and Mapping Operations
GITA   Geospatial Information Technology Association
Glonass   Russian PNT system similar to the US GPS
GMO   Geospatial Management Office
GMT   Greenwich Mean Time (ZULU Time)
GOCO   Government-owned, contractor-operated
GOES   Geostationary Operational Environmental Satellite (NOAA)
GOS   Geospatial One-Stop
GPS   The U.S. NAVSTAR global positioning system
GTS   Global telecommunications system of the WMO
HALE   High altitude long endurance aircraft or drones
HAZMAT   Hazardous materials
HC   U.N. humanitarian coordinator
HCT   U.N. humanitarian country team
HRPT   High resolution picture transmission
HSIN   Homeland Security Information Network
HSOC   Homeland Security Operations Center
HSPD   Homeland Security Presidential Directive
IAEM   International Association of Emergency Managers
IAP   Incident action plan
IC   Incident commander
ICP   Incident command post
ICRC   International Committee of the Red Cross
ICS   Incident command system
Incident   An event, natural or human-caused, that requires an emergency response
INMARSAT   International Maritime Satellite
INPE   Instituto Nacional de Pesquisas Espaciais in Brazil
INTELSAT   International Telecommunications satellite
IP   Internet Protocol
IPoS   Internet protocol over satellite data format
IR   Infrared portion of the electromagnetic spectrum
ISO   International Organization for Standardization
IT   Information technology

JCC Joint Coordination Center
JFO Joint Field Office
JIC Joint Information Center
JOC Joint Operations Center
LANDSAT NOAA Land satellite
LEO Low Earth Orbit
LiDAR Light detection and ranging
LSA Logistics staging area
MAA Mutual aid agreement
MAC Mapping and Analysis Center
MACC Multi-Agency Coordination Center
MEO Medium Earth orbit
Mitigation Activities designed to reduce or eliminate risks or to lessen the potential effects of an incident
MODIS Moderate Resolution Imaging Spectroradiometer instrument of NOAA
MSS Mobile Satellite Services
MUAV Micro unmanned aerial vehicles
NAD National Asset Database
NASA National Aeronautics and Space Administration
NATO North Atlantic Treaty Organization
NAWAS National Warning System
NBC Nuclear, biological, chemical
NEMA National Emergency Managers Association
NESDIS National Environmental Satellite, Data, and Information Service, NOAA
NGO National Geospatial-Intelligence Agency
NGO Nongovernmental organization
NGS National Geodetic Survey
NIEM National Information Exchange Model
NIMS National Incident Management System
NIP National Infrastructure Protection Plan
NOAA National Oceanic and Atmospheric Administration
NRC National Research Council
NRCC National Response Coordination Center
NRP National Response Plan
NSDI National Spatial Data Infrastructure
NSGIC National States Geographic Information Council
OGC Open Geospatial Consortium
OS Open Source
OSC On-scene commander
PAO Public affairs officer
PDA Preliminary damage assessment
PIO Public information officer
Pixel Picture element; defines the spatial resolution of an image or sensor, the smallest area on the ground

Appendix A: Key Terms and Acronyms

PNT Satellite precision navigation and timing system such as GPS
POES Polar Operational Environmental Satellites of NOAA
POTS Plain old telephone system
RADAR Radio detection and ranging
RCC Regional Response Coordination Center
RCRC Red Cross Red Crescent
Remote sensing Acquiring data from a target without being in physical contact
RF Radio frequency
ROC Regional Operations Center, FEMA
RSS Rich Site Summary – An XML-based format to distribute updated web content
SAR Search and rescue, or synthetic aperture radar
SDSS Spatial decision support system
SEOC State Emergency Operations Center
SITREP Situation report
SLOSH Sea, lake, and overland surges from hurricanes
SMS Short message service text messaging on phones and the web
SOP Standard operating procedure
SRTM Shuttle radar tomography mission
SST Sea surface temperature
TIR Thermal infrared portion of the electromagnetic spectrum
TRMM Tropical rainfall measuring mission of NASA
TX Tabletop exercise
UAV Unmanned aerial vehicle
UN OCHA U.N. Office for Coordination of Humanitarian Affairs
UN SPIDER U.N. Platform for Space-based Information for Disaster Management and Emergency Response
UNITAR United Nations Institute for Training and Research
UNOSAT UN Operational Satellite Applications Program
UPS Uninterruptable power supply
USAR Urban search and rescue
USDOT U.S. Department of Transportation
USGS U.S. Geological Survey
UTL Universal task list
V&TC Voluntary & Technical Community
VBMP Virginia Base Mapping Program
VOAD Voluntary Organizations Active in Disasters
VOSOCC Virtual On Site Operations Coordination Center
VSAT Very small aperture terminal
V-Zone The coastal high hazard area
WAAS Wide area augmentation system for GPS in the United States
WHO World Health Organization
WMD Weapons of mass destruction
WMO World Meteorological Organization
XML Extensible markup language

Appendix B: Selected Bibliography

Alexander, D. 2000. *Confronting Catastrophe*. Oxford, U.K.: Oxford University Press. Baker, J. C., B. Lachman, D. Frelinger, K. M. O'Connell, A. C. Hou, M. S. Tseng, D. T. Orletsky, and C. Yost. 2004. *Mapping the Risks: Assessing the Homeland Security Implications of Publicly Available Geospatial Information*. Santa Monica, Calif.: RAND Corporation.

Bruzewicz, A. J. 2003. Remote Sensing Imagery for Emergency Management. Pp. 87–97 in S. L. Cutter, D. B. Richardson, and T. J. Wilbanks (eds.), *The Geographical Dimensions of Terrorism*. New York: Routledge.

Burton, I., R. W. Kates, and G. F. White. 1993. *The Environment as Hazard*, 2nd Edition. New York: Guilford Press

Committee on Planning for Catastrophe: A Blueprint for Improving Geospatial Data, Tools, and Infrastructure, National Research Council. 2007. *Successful Response Starts with a Map: Improving Geospatial Support for Disaster Management* National Academies Press, Washington, D.C. http://www.nap.edu/catalog/11793.html

Cutter, S. L. 2003. GI Science, Disasters, and Emergency Management. *Transactions in GIS* 7(4):439–445.

DHS (Department of Homeland Security). 2005. *Challenges in FEMA's Map Modernization Program*. Washington, D.C.: Department of Homeland Security Office of the Inspector General, OIG-05-44.

ESRI (Environmental Systems Research Institute). 2001. GIS for Homeland Security: An ESRI White Paper. Redlands, Calif. Available at http://www.esri.com/library/whitepapers/pdfs/homeland_security_wp.pdf [accessed May 3, 2013].

Everstine, B. 2013 Drones Play Role in Disaster Response in *Air Force Times* Available at http://www.airforcetimes.com/article/20131126/NEWS/311260021/[accessed May 4, 2013].

FGDC (Federal Geographic Data Committee). 2001. Homeland Security and Geographic Information Systems. Washington, D.C.: Department of the Interior. Available at http://www.fgdc.gov/library/whitepapers-reports/white-papers/homeland-security-gis [accessed May 4, 2013].

Galloway, G. E. 2003. Emergency Preparedness and Response—Lessons Learned from 9/11. Pp. 27–34 in S. L. Cutter, D. B. Richardson, and T. J. Wilbanks (eds.), *The Geographical Dimensions of Terrorism*. New York: Routledge.

Goodchild, M. F. 2003. Geospatial Data in Emergencies. Pp. 99–104 in S. L. Cutter, D. B. Richardson, and T. J. Wilbanks (eds.), *The Geographical Dimensions of Terrorism*. New York: Routledge.

Greene, R. H. 2002. *Confronting Catastrophe: A GIS Handbook*. Redlands, Calif.: ESRI Press. Haddow, G. D., and J. A. Bullock. 2003. *Introduction to Emergency Management*. Boston, Mass.

Longley, P. A., M. F. Goodchild, D. J. Maguire, and D. W. Rhind. 2005. *Geographic Information Systems and Science*, 2nd Edition. New York: Wiley.

NAPA (National Academy of Public Administration). 1998. *Geographic Information for the 21st Century: Building a Strategy for the Nation*. Washington, D.C.: NAPA.

Nayak, S. and Zlatanova S. 2008. Remote Sensing and GIS Technologies for Monitoring and Prediction of Disasters Springer-Verlag Press, Berlin Heidelberg 271pp.

National Commission on Terrorist Attacks Upon the United States. 2004. *The 9/11 Commission Report: Final Report of the National Commission on Terrorist Attacks upon the United States*. New York: W.W. Norton. Available at http://www.gpoaccess.gov/911/ [accessed May 4, 2013].

National Governors' Association. 1979. *1979 Emergency Preparedness Project: Final Report*. Washington, D.C.: National Governors' Association Office of State Services.

NRC (National Research Council). 1993. *Toward a Coordinated Spatial Data Infrastructure for the Nation*. Washington, D.C.: National Academy Press.

NRC. 1994. *Promoting the National Spatial Data Infrastructure Through Partnerships*. Washing- ton, D.C.: National Academy Press.

NRC. 2002. *Making the Nation Safer*. Washington, D.C.: National Academy Press. NRC. 2003. *IT Roadmap to a Geospatial Future*. Washington, D.C.: National Academies Press.

NRC. 2005. *Summary of a Workshop on Using Information Technology to Enhance Disaster Management*. Washington, D.C.: National Academies Press.

NRC. 2006. *Facing Hazards and Disasters: Understanding Human Dimensions*. Washington, D.C.: National Academies Press.

Oshman, Y., and M. Isakow. 1999. Mini-UAV Attitude Estimation Using an Inertially Stabilized Payload. *IEEE Transactions on Aerospace and Electronic Systems* 35(4):1191–1203.

Perham, S. D. 2009, *GIS Mapping and Data-Management as an Information-Sharing Model to Promote the Coordination of Humanitarian Programmes*, Royal Roads University

Thomas, D. S. K., S. L. Cutter, M. E. Hodgson, M. Gutekunst, and S. Jones. 2003. Use of Spatial Data and Geographic Technologies in Response to the September 11 Terrorist Attack on the World Trade Center. Pp. 147–162 in *Beyond*

September 11th: An Account of Post-disaster Research. Special Publication 39. Boulder, Colo.: Natural Hazards Research and Applications Information Center, University of Colorado.

Waugh, W. L., Jr. 1988. Current Policy and Implementation Issues in Disaster Preparedness. Pp. 111–125 in L. K. Comfort (ed.), *Managing Disaster: Strategies and Policy Perspectives.* Durham, N.C.: Duke University Press.

Waugh, W. L., Jr. 2000. *Living with Hazards, Dealing with Disasters: An Introduction to Emergency Management.* Armonk, N.Y.: M. E. Sharpe.

Zerger, A. and Smith, D. I. Impediments to Using GIS for Real-Time Disaster Decision Support Computers, *Environment and Urban Systems* Volume 27, Issue 2, March 2003, Pages 123–141

Appendix C: Selected Websites

Asia Disaster Reduction Center (ADRC)
  http://www.adrc.asia
Crisis Mappers
  http://crisismappers.net
Disaster Risk Reduction, Climate Change Adaptation and Human Security
  http://www.adpc.net/DDRCCA/
Disaster Risk Reduction Project Portal
  http://www.drrprojects.net/drrp/drrpp/home
ESRI (ArcGIS) Emergency Management portal
  http://www.esri.com/services/disaster-response/disaster-relief
FEMA Emergency Management Institute (online training and materials)
  http://training.fema.gov/is/
Field Guide to Humanitarian Mapping by Map Action
  http://www.mapaction.org/?option=com_mapcat&view=mapdetail&id=2426
GEOSS Best Practices Wiki
  http://www.princidentionweb.net
GISCorps
  http://www.giscorps.org
Google Crisis response
  http://www.google.org/crisisresponse/resources.html
  http://www.preventionweb.net/files/27023_guidebookenwatermarkedred.pdf
Humanitarian Dashboard
  https://assessments.humanitarianresponse.info/humanitarian-dashboards
Humanitarian OpenStreetMap Team
  http://hot.openstreetmap.org
ICRC The International Committee of the Red Cross
  http://www.icrc.org/eng/

INaSAFE
    http://inasafe.readthedocs.org/en/latest/contents.html
International Charter for Space and Major Disasters
    http://www.disasterscharter.org/home
Landsat
    http://landsat.usgs.gov
Mapmill disaster assessment using crowd sourcing
    http://opengov.newschallenge.org/open/open-government/submission/mapmill-crowdsourced-disaster-damage-assessment/
Multi-language Glossary on Natural Disasters
    http://glossary.adrc-web.net/trans2.asp?lang=en
NOAA GOES Server
    http://www.goes.noaa.gov
NOAA POES
    http://poes.gsfc.nasa.gov
PreventionWeb
    http://www.princidentionweb.net
QGIS Open Source GIS software
    http://qgis.org
ReliefWeb
    http://www.reliefweb.int
The Role of Remote Sensing in Disaster Management
    http://www.iclr.org/images/Role_of_Remote_Sensing_in_Disaster_Management.pdf
UN Disaster GIS Icons
    http://reliefweb.int/sites/reliefweb.int/files/resources/2012_OCHA_humanitarian_icon_editablepdf.pdf
UN SPIDER Knowledge Portal
    http://www.un-spider.org
UN SPIDER Links and Resources
    http://www.un-spider.org/links-and-resources
UN Office of Outer Space Affairs (OOSA)
    http://www.oosa.unvienna.org
World Bank Disaster Risk Management
    http://web.worldbank.org/WBSITE/EXTERNAL/TOPICS/EXTURBANDEVELOPMENT/EXTDISMGMT/0,,contentMDK:20857119~menuPK:2337111~pagePK:210058~piPK:210062~theSitePK:341015,00.html

About the Author

Scott Madry, Ph.D., is a specialist in remote sensing and GIS for regional environmental and cultural applications, including disaster management. He is a member of the faculty of the International Space University in Strasbourg, France, and is a research associate professor of archeology at the University of North Carolina at Chapel Hill. He has been involved with the International Space University since 1987 and has taught in over 25 ISU summer programs around the world. He was on the resident ISU faculty in Strasbourg, France, for 3 years and is currently the Program Director of the Southern Hemisphere Summer Space Program, held most recently in Adelaide, Australia. He is the founder and president of Informatics International, Inc., an international remote sensing and GIS consulting company located in Chapel Hill, NC. He specializes in natural and cultural resource management and disaster applications. He received his Ph.D. in 1986 at UNC-CH and worked at the Institute for Technology Development Space Remote Sensing Center at NASA for 3 years, then taught at Rutgers University for 9 years before teaching at ISU and moving back to UNC. His research includes remote sensing, GIS and GPS applications, and he has done work in North America, Africa, and Europe and was the co-editor-in-chief of the 1,200-page, two-volume *Handbook of Satellite Applications* for Springer in 2012. He has given over 150 short courses and educational programs in 30 countries around the world and is very involved in Open Source GIS, plus has taught QGIS/GRASS courses around the world. He is an advanced disaster instructor with the American Red Cross and has deployed to multiple disasters with them, mostly as an Emergency Operations Center liaison, but has worked in several other capacities. He is also active with the GISCorps in providing remote data analysis in disasters, and he was awarded, along with other GISCorps volunteers, the 2012 President's Volunteer Service Award for 2012 by President Barack Obama.

MIX
Papier aus verantwortungsvollen Quellen
Paper from responsible sources
FSC® C105338

If you have any concerns about our products,
you can contact us on
ProductSafety@springernature.com

In case Publisher is established outside the EU,
the EU authorized representative is:
Springer Nature Customer Service Center GmbH
Europaplatz 3, 69115 Heidelberg, Germany

Printed by Libri Plureos GmbH
in Hamburg, Germany